MINERALS AND ROCKS
in colour

by J. F. KIRKALDY, D. Sc.

Emeritus Professor of Geology
formerly Head of the Department of Geology
Queen Mary College, London.

photographs by

FOLKE JOHANSSON

and

MICHAEL ALLMAN

BLANDFORD PRESS

POOLE   DORSET

First published in the English edition 1963
© Copyright by Blandford Press Ltd.,
Link House, West Street / Poole, Dorset BH15 1LL
2nd impression November 1964
3rd impression December 1965
2nd edition January 1968
reprinted January 1970
reprinted January 1972
reprinted February 1973
3rd edition 1976

ISBN 07137 0783 6

*This was originally published in Sweden
entitled* Stenar i färg
by Almqvist and Wiksell. Stockholm

Printed in Holland by Yselpress, Deventer
and bound by Richard Clay (The Chaucer Press) Ltd.

CONTENTS

I	MINERALS AND ROCKS	81
	The Layered Nature of the Earth	81
	The Continental Masses	83
	The Nature and Origin of the Three Main Groups of Rocks	86
	The Geological Time Scale	89
	The Origin of Mineral Deposits	91
II	IDENTIFICATION OF MINERALS	96
	Table for Identification of Minerals	103
III	DESCRIPTION OF THE COMMONER MINERALS	117
	Native Elements	117
	Sulphides	119
	Oxides	123
	Halides	128
	Carbonates	130
	Sulphates	132
	Wolframites	133
	Silicates	133
	Radioactive Minerals	144
IV	THE IDENTIFICATION OF ROCKS	146
V	DESCRIPTION OF THE COMMONER ROCKS	150
	Igneous	150
	Sedimentary	156
	Metamorphic	169
	Meteorites	173
VI	FOR FURTHER READING	174
VII	GLOSSARY	175
	Index	181

PREFACE

The colour photographs of minerals and rocks are the work of Folke Johansson. They were taken to illustrate a book, Stenar i färg (Stones in Colour), by Per H. Lundegårdh, published in Stockholm in 1960. Mr. Lundegårdh selected the specimens to be photographed. The majority were in the collections of the Mineralogical Department of the Swedish Riksmuseet or of the Swedish Geological Survey. Further specimens were lent by the Danish Geological Survey and the Mineralogical Museum, Copenhagen, by the Finnish Geological Survey, by the Mineralogical Museum and Institute of Geology at Oslo, and by the Geological Institute at Bergen in Norway.

For this English edition, Mr. Johansson's magnificent colour photographs have been retained. The text has been completely rewritten and illustrated by new line drawings. For the original Swedish edition, Mr. Lundegårdh had naturally concentrated on those minerals and rocks which are to be found in Scandinavia, but in other parts of the world, particularly in the British Isles, there occur a number of common minerals and rocks, which are either extremely rare or absent in Scandinavia. This applies particularly to the sedimentary rocks (limestone, coal, etc.) Some of the most characteristic Scandinavian rock-types, such as those shown in Plates 33 and 43, scarcely occur in the natural state outside Scandinavia, though they have been exported widely for use as decorative stones. It therefore follows that not all the minerals and rocks mentioned in the text are illustrated in colour. It is, however, hoped that sufficient information has been given for their identification.

It has been assumed that the users of this book will be anxious not only to identify minerals and rocks, but will be curious as to their mode of formation and their economic uses, so these aspects are discussed as fully as space permits. It is hoped that the user will be encouraged not merely to collect specimens of attractive looking minerals and rocks, but also to examine in the field their relations to one another and to the country rocks and hence be lead to study more fully that most fascinating of the sciences – Geology.

The characters given for identification have been limited to those that can be seen with the naked eye or with a hand lens. Space does not permit adequate treatment of the features that can be seen in thin-section under the petrological microscope, but in discussing the origin of certain rocks reference has had to be made to their micro-scopical characters. This aspect can be pursued, if desired, in the books listed under Section VI.

The author would like to express his gratitude to his colleague Dr. A. C. Bishop, for his critical and constructive reading of the typescript, to Mr. C.G. Jay for the majority of the line drawings and to the publishers for their help throughout and especially for arranging for the fine colour photographs to become available to English speaking geologists.

Geology Department
Queen Mary College
University of London

J. F. Kirkaldy
May, 1963

The call for a new edition has enabled minor changes and additions to be made to the text, whilst eight new pages of colour photographs have been added. The new colour photographs (268-290) have all been taken by Mr. M. A. Allman, who was also responsible for the colour photography in the author's companion volume, *Fossils in Colour*. The specimens from the collections of the Geology Department, Queen Mary College, were selected to illustrate a number of minerals and wellknown rock types, which are not found in Scandinavia and therefore were not illustrated in the original Swedish edition.

J. F. Kirkaldy
November, 1967

1. Gold nugget
2. Flakes of gold in quartzite
3. Silver
4. Copper and lead
5. Kohinoor Diamond after its latest cutting
6. Diamond in 'blue ground', Kimberley, South Africa
7. Graphite

9

21. Blende (sphalerite)
22. Massive form of blende (sphalerite)
23. Molybdenite (scales in iron-stained quartz)
24. Molybdenite in quartz
25. Stibnite (antimony glance) in quartz

26. Crystal of cobaltite
27. Rock crystal (transparent quartz)
28. Smoky quartz
29. Milky quartz (the common form of quartz)
30. Smoky quartz
31. Massive smoky quartz
32. Amethystine quartz

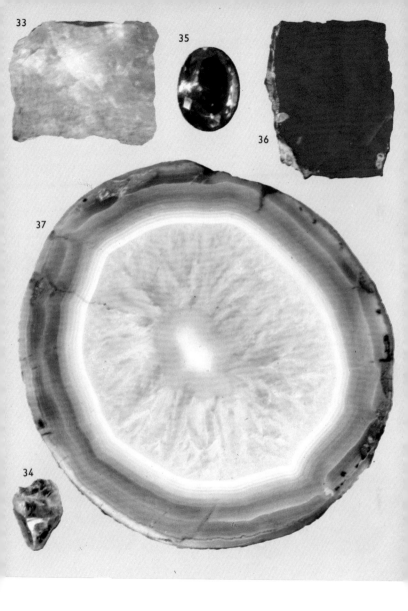

44. Flint nodule from the Chalk
45. Rutile in quartz
46. Cassiterite (tin stone)

47. Corundum (red)
48. Ruby, cut
49. Sapphire, cut

50. Polished specular hematite
51. Banded 'bloodstone ore' carrying specular hematite, Sweden.
52. 'Bloodstone ore' with crystals of magnetite, Sweden.
53. Crystal of magnetite
54. Magnetite

17

55

56

57

58

59

66. Cryolite (white) with chalybite (brown), galena and copper pyrites, Greenland

67. Rhombs of calcite, (the transparent variety, 'iceland spar', shows double refraction).

68. Twinned crystal of calcite

69. Crystal of calcite

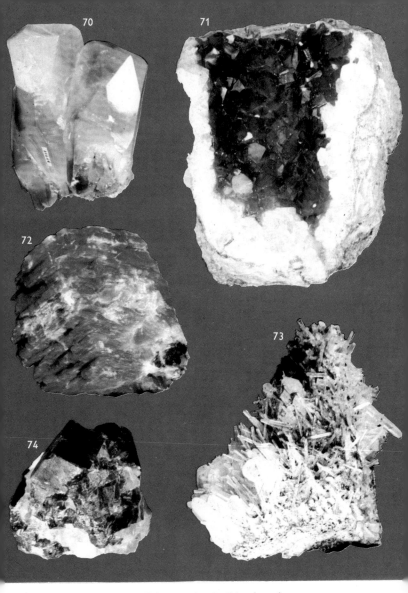

70. Calcite, variety 'nail-head spar'
71. Calcite, brown stained
72. Calcite, stained with iron oxide
73. Aragonite
74. Chalybite (siderite)

88. Plagioclase feldspar (oligoclase)
89. Plagioclase feldspar (labradorite), polished
90. Leucite
91. Nepheline crystals in basalt
92. Scapolite
93. Garnet, variety almandine
94. Garnet, variety melanite
95. Garnet, polished
96. Garnets in quartz

97. Olivine
98. Idocrase (vesuvianite)
99. Zircon
100. Topaz (top crystal)
101. Topaz in quartz

102. Andalusite
103. Sillimanite
104. Kyanite
105. Epidote with stringers of iron ore
106. Epidote in iron ore

107. Enstatite
108. Diopside
109. Augite

110. Chrome-diopside
111. Rhodonite
112. Hornblende

113. Hornblende
114. Anthophyllite
115. Tremolite
116. Actinolite
117. Asbestos, fibrous amphibole

118. Asbestos, fibrous amphibole
119. Aquamarine, cut
120. Emerald, cut
121. Beryl in quartz
122. Emerald in micaceous rock

123. Beryl
124. Beryl, white variety

125.
126. } Tourmaline in quartz

127. Black tourmaline (schorl) in quartz

128. Pink tourmaline (rubellite)
129. Cordierite
130. Cordierite in feldspar
131. Staurolite in mica-schist
132. Apophyllite
133. Natrolite

134. Heulandite in amygdales in bas-
     altic lava
135. Muscovite (white mica)

136. Lepidolite (lithia mica), pink
     flakes in feldspar
137. Flakes of lepidolite in pegmatite

138. Biotite (dark mica), with feldspar
139. Biotite
140. Fuchsite (chromium mica)

141. Fuchsite (chromium mica)
142. Chlorite
143. Serpentine

144. Serpentine
145. Talc
146. Glauconite

147. Brown sphene (titanite) in quartz and feldspar
148. Amber

149. Amber
150. Pitchblende
151. Uraninite
152. Monazite
153. Crystal of Steenstrupine, Greenland
154. Euxenite with feldspar
155. Euxenite

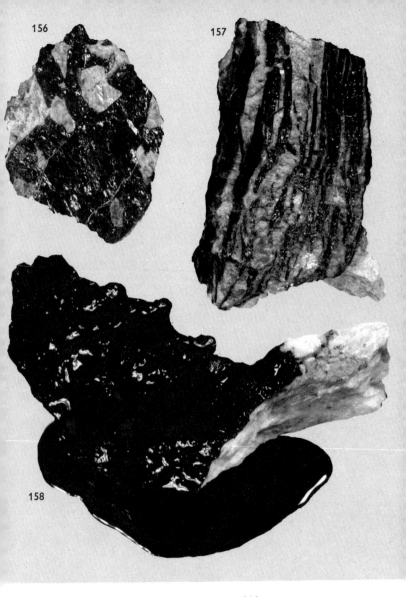

156. Gadolinite with feldspar
157. Allanite (orthite)
158. Asphalt

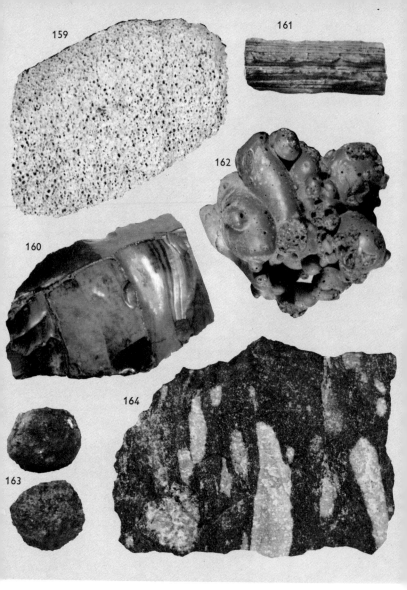

159. Pumice, Iceland
160. Obsidian, Iceland
161. Ropy Lava, Iceland

162. Oxidized upper surface of Lava
Flow, Iceland
163. Lapilli (volcanic bombs)
Iceland
164. Agglomerate, Sweden

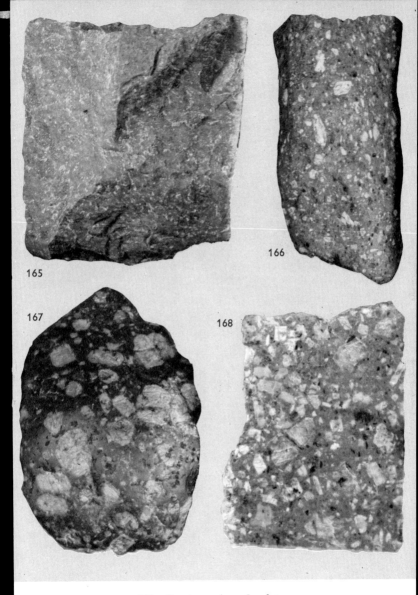

165. Quartz-porphyry, Sweden

166. ⎫

166. ⎬ Varieties of coarse-grained

167. ⎨

168. ⎭ Scandinavian porphyries

169. 170. } Fine-grained porphyry, Sweden

171. Flow-banded porphyry, Sweden

172. Porphyritic granite, Sweden

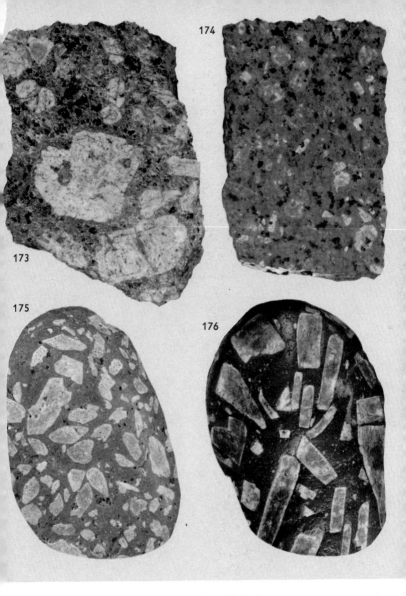

174

173

175

176

173. Rapakivi-porphyry, Finland
174. Pebble of rapakivi-porphyry, Finland
175. } Pebbles of rhomb-porphyry,
176. } Denmark

41

177. Rhomb-porphyry, Norway
178. Rhomb-porphyry conglomerate,
     Norway
179. Alnoite, Sweden

180. Plagioclase-porphyry, Sweden
181. Uralite-porphyry, Sweden
182. Andesite, Sweden

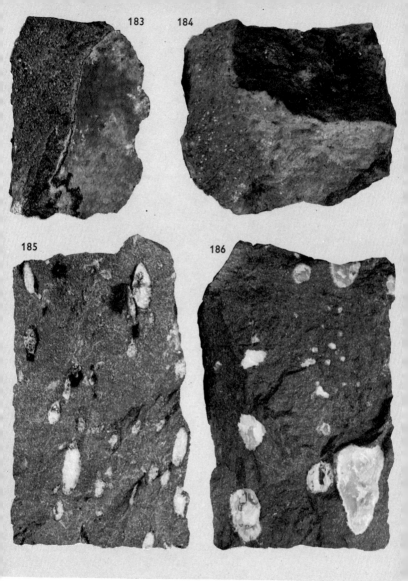

183.   Andesite, Sweden
184.   Basalt, Sweden
185. } Vesicular basalt, 'Almond-stone',
186. } Sweden

187.
188. } Varieties of Swedish diabases
189. } (dolerites)
190. Hyperite, Sweden

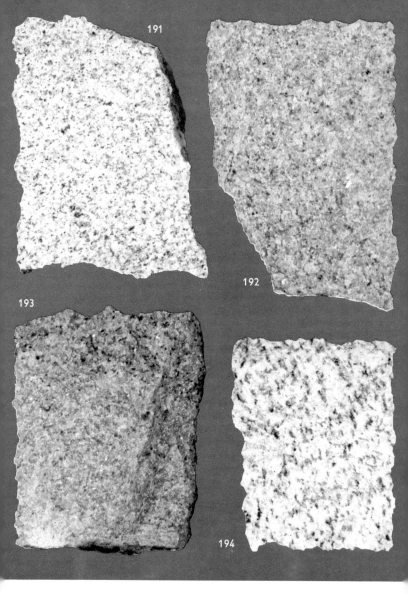

191. White microgranite, Sweden
192. Red microgranite, Sweden
193. Red microgranite cut by aplite vein, Sweden
194. Grey fine-textured granite, Finland

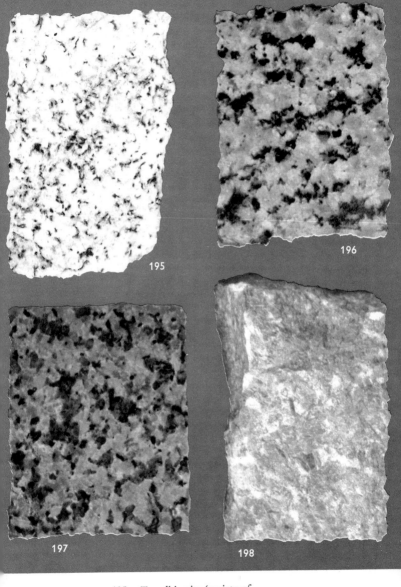

195. Trondhjemite (variety of
      granodiorite), Norway
196. } Red medium-textured granites,
197. } Sweden
198. Coarse-textured granite, Sweden

47

199.
200. } Varieties of coarse-textured gra-
201. } nites from Scandinavia.
202.

203

204

205

203. ⎫
204. ⎬ Porphyritic granites, Denmark

205. Syenite, Island of Bornholm, Sweden

49

206. Porphyritic rapakivi-granite,
     Finland
207. Nordmarkite, quartz-syenite,
     Norway

208. Laurvikite, light variety,
     Norway
209. Laurvikite, dark variety,
     Norway

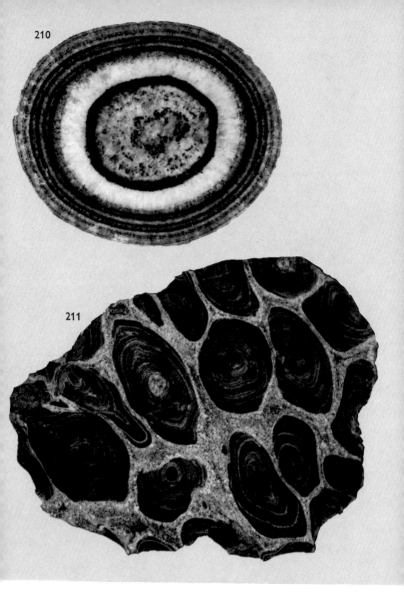

210. 'Orb' from orbicular granite, Finland

211. Orbicular granite, Finland (Scale 1:2)

212. Aplite vein in gneiss, Island of
     Bornholm, Denmark
213. Graphic granite, Sweden
214. Pegmatite with beryl, Norway.

215. Pegmatite, Sweden
216. Quartz-diorite or granodiorite, Sweden
217. Norite, Sweden

218. Olivine-gabbro, Sweden
219. Olivine-gabbro (allivalite),
     Sweden
220. Anorthosite, Sweden

221. Anorthosite, Norway
222. Peridotite, Sweden
223. Tillite, Norway

224. Polygenetic conglomerate,
Norway (Scale 1:2)
225. Conglomeratic sparagmite,
Norway
226. Bedded sandstone, Sweden

227. Sandstone, Sweden
228. Medium-grained sandstone,
     Island of Bornholm, Denmark
229. Ripple-marked sandstone,
     Sweden (scale 1:4$^1/_2$)

230. Shale, Sweden
231. Partly burnt Alum Shale,
     Sweden
232. Grey Orthoceras limestone,
     Sweden

59

233. Orthoceras limestone, Sweden
234. Orthoceras limestone with
     Orthoceras, Sweden
235. Polished limestone, 'marble',
     Faxe, Denmark

236. Coral limestone, Faxe,
     Denmark
237. Dolomite concretion, Faxe,
     Denmark
238. Brown Coal (lignite), Denmark

239. Leptite (granulite) in white marble, Sweden
240. Lineated amphibolite, Sweden
241. Serpentinite, Finland

242. Granite-gneiss, Sweden
243. Augen-gneiss, Sweden

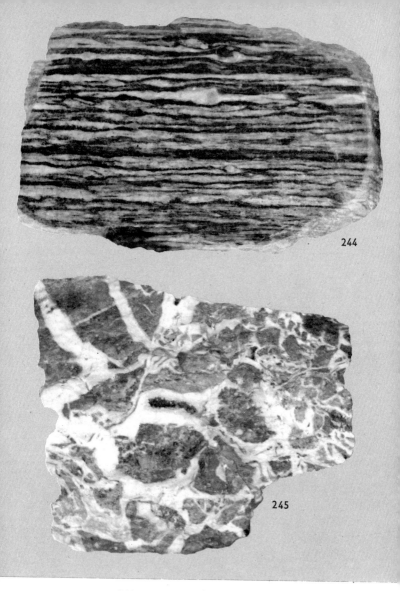

244. Mylonitic gneiss, Sweden
245. Quartz-veined breccia, Sweden

246. Deformed conglomerate,
     Bergen, Norway, Scale 1:2

247.
248. } Quartzite, Sweden.

249. Mica-schist, Norway
250. Phyllite, Norway

251

252

253

254

251. Green slate, Norway
252. Hornfels with quartz vein, Norway

253. Marble, 'Rouge Griotte', Belgium
254. Marble, Norway.

255.
256. } Varieties of gneiss, Sweden
257.

258. Banded gneiss, Sweden
259. Strongly deformed gneiss,
     Finland

260. Folded gneiss, Norway
261. Garnetiferous gneiss, Sweden

262. Migmatite, Sweden
263. Eclogite, Norway

264. Garnetiferous anorthosite-gabbro, Norway, Scale 1:2
265. Iron meteorite, Sweden
266. Polished iron meteorite, Sweden
267. Stony meteorites, Sweden

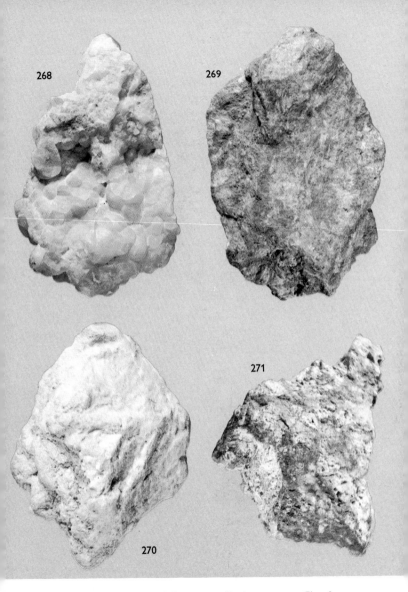

268. Sulphur – 269. Realgar – 270. Orpiment – 271. Cinnabar

272. Marcasite nodule – 273. Kidney iron ore (hematite) – 274. Pyrolusite (fibrous habit) – 275. Pyrolusite (dendritic habit)

276. Lapis-Lazuli – 277. Spinels – 278. Dolomite (brown) and quartz (white) –
279. Garnets in skarn

280. Sphaerulitic rhyolite, Channel Isles – 281. Serpentinite, England

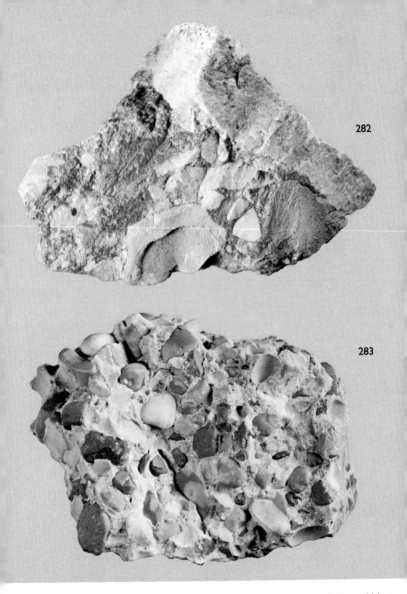

282

283

282. Breccia of limestone fragments, England – 283. Conglomerate of flint pebbles, 'Hertfordshire Puddingstone', England

284

285

284. Bone Bed of fossil fish, England – 285. Polished gastropod limestone, 'Purbeck Marble', England

I. MINERALS AND ROCKS

The earth's crust is made up of minerals and rocks. Minerals are inorganic substances with a definite chemical composition. Each mineral is composed of certain elements present in fixed proportions. For example, quartz is silicon dioxide with silicon (Si) and oxygen (O) combined in the proportions of $1:2$, so its chemical formula is SiO_2. Orthoclase feldspar is more complex. It is a silicate mineral containing potassium (K) and aluminium (Al) in the proportions shown by its formula $K\,Al\,Si_3O_8$. In some minerals, however, there may be slight variation in the proportions of the various elements. Biotite (dark mica) is another complex silicate of magnesium (Mg), aluminium (Al), potassium (K), hydrogen (H) and iron (Fe) and has the 'general' formula $(HK)_2$, $(MgFe)_2$, $(AlFe)_2$, $(SiO_4)_3$.

The atoms (the single particles of each element) of a mineral are arranged in a definite pattern, the atomic structure. This is also a characteristic of the mineral and determines many of the distinctive properties (shape, weight, ease of splitting, etc.) by which each mineral can be identified.

Rocks are much more variable in composition than minerals. Many rocks are composed of a number of minerals. For example, the well known Cornish granite can be seen by the unaided eye to be made up very largely of three interlocking minerals, clear quartz, white orthoclase and dark shining biotite. Clay is also made of minerals, but these are so minute that they are only visible under the electron microscope magnifying many thousands of times. Other rocks, such as coal and chalk, are composed largely of material of organic origin.

We have used the terms minerals and rocks in their strictly geological sense and shall continue to do so throughout this book. The 'popular' idea of rocks and minerals is at times rather different. A rock is commonly thought of as something that must be fairly tough and compact. But if one regards a rock as a constituent of the earth's crust, then granite is simply a hard rock, clay a soft rock. In the same way, minerals are commonly regarded as having an economic value. In economic circles, the term mineral is therefore extended to include many substances e.g., coal, glass sand, chalk, petroleum, etc., which are geologically speaking rocks.

The Layered Nature of the Earth

The earth is spheroid, 6,400 kilometres (4,000 miles) in radius. The specific gravity of the earth as a whole is 5.5, that is, it weighs 5.5 times as much as would the same volume of water. But the great majority of rocks and minerals which are to be found on the surface of the earth have densities differing but little from 3. The deeper parts of the earth must therefore be composed of material of considerably greater density.

From a study of the path taken by, and the differing velocities of earthquake waves, it is possible to infer much about the make up of the deeper parts of the earth. The presence of a number of concentric surfaces of discontinuity have been recognized. At

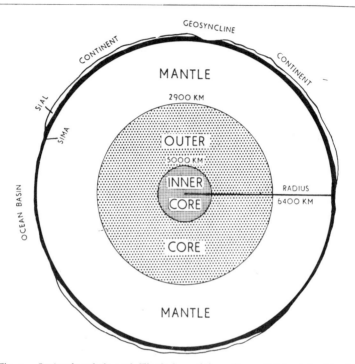

Fig. 1. *Section through the earth. The thickness of the earth's crust (sial and sima) has had to be greatly exaggerated.*

each surface there are considerable changes in the physical properties of the layers in contact. A major change at a depth of 2,900 kilometers, separates the core from the mantle (Fig. 1). The material that composes the core must be at very high pressure and very high temperature, perhaps 5,000° C. Its density is estimated to increase from 9.9 at the outer core to 12 or more towards the centre of the earth. Naturally we have no direct knowledge of the nature of the material that composes the core. It may be nickel-iron or may be wholly, or in part made up of enormously compressed gases.

The mantle surrounding the core is also layered, with the layers decreasing in density outwards. The outermost layer comprises the crust of the earth. There are major differences between the crust underlying the oceans and the continents. Continental crust has an average density of 2.85 and its thickness varies from 25 to 70 kilometres. The thickest parts underlie the highest mountain ranges. It is composed of rocks rich in silicon (Si) and aluminium (Al). Oceanic crust is formed by heavier rocks, rich in silicon (Si) and Magnesium (Mg). It's density is greater than 3 and it is thinner, on average about 9 kilometres.

The Continental Masses

The rocks that make up the continental areas are also layered. One can see this very clearly by examining the sections exposed in cliffs, quarries, mountain sides and elsewhere. In some places,

the rock layers are either horizontal or gently inclined (dipping) at a low angle. In other places they are most violently contorted. The strongly folded rocks occur in narrow belts separated by wider areas of flat lying rocks. The rocks of the high mountain chains, the Alps, Himalayas, Andes, the Rockies, etc., are very strongly folded. These are geologically speaking the 'young' mountain chains. In other parts of the world, in areas of considerably lower relief, such as Scandi-

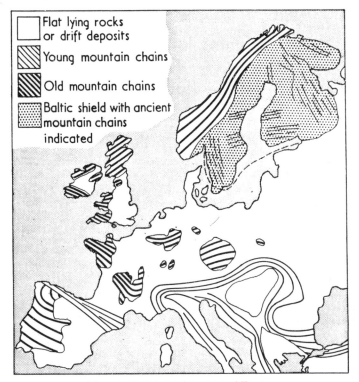

Fig. 2. *Simplified geological map of Europe.*

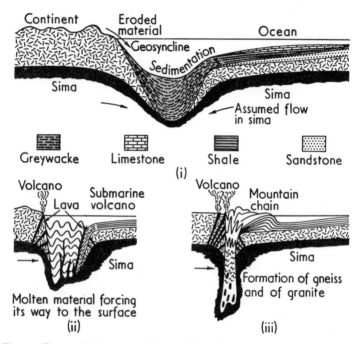

Fig. 3. *The growth of a mountain chain. (i) The infilling of the geosyncline with sediment, (ii) The first stages of folding and accompanying vulcancity, (iii) Climax of folding, followed by uplift.*

navia and the Scottish Highlands, similarly strongly folded rocks occur. These are the 'old' mountain chains. In the remote geological past, these areas must have been high mountain chains, but they have exposed for so long to erosion, that is the wearing away of rocks by the action of rain, ice, etc., that great quantities of rock material have been removed and their relief much reduced. In the 'shield areas' of the world, much of Canada, Finland, etc., the relief is usually extremely low. A thin layer of un-

consolidated sands, gravels and clays, covers extremely contorted solid rocks. These 'drift' deposits were laid down by the ice sheets which covered these areas as recently as a few tens of thousands of years ago. The greatly folded solid rocks are the deeply eroded parts of very ancient mountain chains. The young mountain chains were formed a few tens of millions of years ago, the old mountain chains of Scotland and Norway, 400 or more million years ago, the ancient mountain chains of the shield areas, a thousand or more million years ago. (Fig. 2).

These mountain chains all originated in *geosynclines,* that is in narrow elongated belts of the earth's crust which subsided as shown in Fig. 3 (i). The material eroded from the surrounding land areas was transported to and deposited in the geosynclines to form new rocks. Eventually subsidence ceased and the sides of the geosyncline moved towards each other, causing great compression. At the same time fissures were formed, up which rose molten rock material from the sima and deeper parts of the sial (Fig. 3 (ii)). This molten material (magma), highly charged with gases under high pressure, sometimes reached the earth's surface to escape through volcanoes, at other times it was frozen at a depth of a few miles to form great masses of granite.

The chain of partly volcanic islands extending from Sumatra, through Java towards New Guinea are believed to be centred on a developing geosyncline. In the young mountain chains we can study the rocks and structures developed in the highest parts of the folded belt, in the older mountain chains those formed at greater depth, whilst in the shield areas erosion has cut down to expose the roots of the former mountain chains.

Whilst at intervals during the past, certain parts of the continents have been affected by great compression, at other times some areas have been subjected to tension. Eventually the rocks fractured along narrow fault zones to produce fault bounded blocks (Fig. 4). Faults are lines of weakness up which material at high temperature can rise to form dykes or mineral veins or even to spread out on the surface as great sheets of lava. At the Giant's Causeway and elsewhere in Antrim or, in Iceland, in the Snake River area in Idaho and Nevada, can be studied the lava fields formed by such fissure eruptions.

In those parts of the land areas, such as southern and eastern England and the central United States, which have not been affected during the past few hundred million years by strong folding movements or by major faulting, the rocks occur as either horizontal or gently dipping layers. Locally they may have been folded into either anticlines or synclines (Fig. 5), but in general the folding is very insignificant compared with that to be seen in the mountain chains.

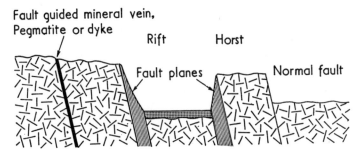

Fig. 4. *Some of the effects of faulting.*

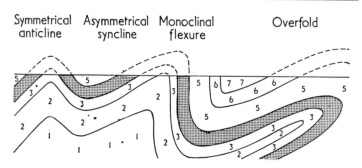

Symmetrical anticline   Asymmetrical syncline   Monoclinal flexure   Overfold

Fig. 5. *The effect of folding increases in intensity towards the right. The beds are numbered in the order in which they were deposited. Note that the beds on the lower limb of the overfold or recumbent fold have been inverted.*

The Nature and Origin of the Three Main Groups of Rocks

The rocks of the earth's crust fall into three groups—Igneous, Sedimentary and Metamorphic.

The igneous rocks are crystalline for they have originated from the consolidation of magma. They are normally made up of tightly interlocking crystals of a variety of minerals. If the magma cooled slowly, the individual crystals were able to grow to a large size, perhaps several inches in length, and a coarse-grained rock, such as granite was produced. With more rapid cooling, finer-grained rocks were formed. Very rapid chilling, such as of the upper surface of a lava flow, produced a volcanic glass (obsidian) with the material chilled too quickly for crystallization to occur.

Some igneous rocks have a porphyritic texture (See pp. 41 and 49) with large crystals (phenocrysts) of one mineral, set in a finer-grained groundmass of other minerals. Such rocks must have had a complex cooling history. First a stage of slow cooling

with the growth of the large phenocrysts. Then more rapid cooling with the quick crystallization of the groundmass. This second period of cooling was often due to the partly crystalline magma being moved bodily, perhaps during folding movements, nearer to the earth's surface or at least into a region of lower temperature.

Magmas crystallize at temperatures between 1,200°C and 600°C. The first minerals to form are those silicates with a lower proportion of silicic acid and a higher proportion of iron and magnesium. The remaining liquid, therefore, becomes more acid and the later minerals to form have a higher silicon content, until finally quartz crystallizes last of all. The minerals of an igneous rock therefore crystallize in the order of their decreasing basicity. In a granite the feldspars are normally large and well formed, they crystallized first, then came the micas and last of all quartz, which filled in the spaces between the earlier formed minerals and therefore could not form well shaped crystals.

A volcanic cone (Fig. 6) is usually built up of sheets of lava separated by pyroclastic layers. These layers consist of the material thrown out by explosions, caused by the feeder pipe becoming choked, until sufficient gas pressure built up for the blockage to be violently ejected. Such pyroclastic layers are made up of material varying in size from huge boulders (agglomerates) to fine dust (tuff). Sometimes tuffs contain well formed crystals of minerals. These must have formed in the magma chamber and then have been blown high into the air to fall on the flanks of the volcano. The coarsest material fell near the volcano, the finer the material the further away. Fragments were often torn off the sides of the neck of the volcano and blown high into the air. As shown in Fig. 6 a volcanic neck may cut through sedimentary rocks, and, if so, its pyroclastic debris may include pieces of these sediments. Pieces of chalk, and even fossils from the Chalk, have been found in the agglomerates of a fossil volcano on the island of Arran. No chalk is exposed on the island, but the Chalk must be present at depth.

Acid magma is much more viscous than basic magma, so one finds that much pyroclastic material is usually interbedded with acid lavas, but relatively little with basic lavas.

Sedimentary rocks are formed from the destruction of pre-existing rocks. The stresses produced when rocks expand and contract owing to change of temperature, the effects of the 'frost wedge', the blows of waves or the rasping action of wind and moving water, including ice, break rocks into fragments. In regions of hot climate and heavy rainfall, the soluble constituents of rocks are slowly dissolved

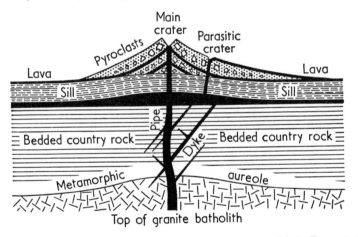

Fig. 6. *Section of a volcano and the associated bodies of igneous rock. Black: Fine-grained igneous rock; Dashes: Coarse-grained igneous rock; Stippled: Pyroclasts with triangles indicating agglomerates; Lined: Bedded sedimentary rock; White: Metamorphic aureole of the granite batholith.*

away. The products of rock destruction (weathering) move downwards under gravity, in part as the solution load of rivers, in part as particles and fragments carried by rivers, by the waves, by the wind or by ice.

During this movement the rock fragments wear away (erode) the surface over which they are moving and are themselves reduced in size and rounded in shape. The transported material is deposited in the hollows of the earth's crust, in the seas and oceans and in the depressions on the land surfaces, to form new sedimentary rocks.

Sedimentary rocks differ from igneous rocks in a number of features. In the first place, they are normally stratified or bedded, occuring as layers, parallel to the surface on which they were deposited. Secondly many sedimentary rocks are clastic, that is they are clearly made up of fragments, either rounded or angular, of pre-existing rocks and minerals. Thirdly sedimentary rocks often contain fossils, the remains of the life of the past. Certain sedimentary rocks such as coal and some varieties of limestone are very obviously composed largely of organic material. Other sedimentary rocks, such as the *Salt Deposits* (the Evaporites) are chemical precipitates, formed by the deposition of the soluble products of weathering. Other sedimentary rocks may contain clastic fragments set in a fine-grained groundmass which may be largely or in part of organic origin or may have been precipitated. Sedimentary material can be considerably modified by post-depositional (diagenetic) changes. It is compacted by the weight of beds deposited later. A freshly-deposited clay contains much water. As this is squeezed out, the clay hardens or lithifies into a mudstone or a shale.

Water moving through porous rocks, may dissolve away calcareous shells at one point and perhaps precipitate the calcium carbonate elsewhere to cement the grains of a sand together to form a fairly hard calcareous sandstone. The character of a sediment may be greatly altered by diagenetic changes. Percolating water may precipitate its dissolved silica to change an originally calcareous rock into a siliceous rock. Minerals such as small cubes of iron pyrites may develop during diagenesis or as concretionary masses or nodules differing in chemical composition from the enveloping rocks.

Metamorphic Rocks are igneous or sedimentary rocks which have been altered by high temperature, strong pressure or by a combination of these. When magma at a temperature of several hundred degrees Centigrade comes in contact with cold rocks, these will be altered. They will be baked, changing their texture and appearance, whilst their mineralogical composition may be altered, owing to the formation of high temperature minerals such as cordierite. The altered rocks are said to be contact or thermally metamorphosed. The width of altered rock, the metamorphic aureole, will depend partly on the temperature difference, but more particularly on the amount of heat exchange which is determined by the volume of the igneous mass. A lava flow will be underlain by a thin selvage, perhaps less than an inch in thickness, of altered rock. A granite batholith, such as that of Dartmoor, hundreds of cubic miles in volume, will be surrounded by an aureole, a mile or more in width. In such a wide aureole one can recognize a zonation as the granite margin is approached. The slates, the 'country rock', into which the granite was intruded, becomes spotted with the development of crystals of cordierite, at first extremely minute, then becoming clearly visible to the naked eye until near to the granite the spotted slates change into tough rock called hornfels in which other minerals are developed.

During periods of mountain building (orogenic episodes) the rocks of geosynclines, whether originally of igneous or sedimentary origin, were regionally metamorphosed. The finer-grained rocks, the clays and the volcanic ashes, were cleaved, with the formation of new planes of parting at high angles to the original bedding. Other rocks suffered drastic reconstitution, with the formation of new minerals, high density minerals, such as garnet and kyanite. These minerals occur in a layered or schistose manner. In the deeper parts of the folded belt, the rocks were permeated by magma, usually acid magma, which had risen from below, to form gneisses. Gneiss differs from schist by showing a much coarser banding of quartzose or feldspathic layers with layers rich in dark coloured silicate minerals. Limestone was recrystallized to form marbles with, of course, the obliteration of any fossils, whilst quartz sands or sandstones were hardened into quartzites. In such regionally metamorphosed areas it is often possible to recognize textural and mineralogical zonation and hence to map the varying intensity of the metamorphic changes.

The Geological Time-Scale

The stratified rocks of the earth's crust have been arranged by geologists in the order in which they were deposited. In any one area one can recognize a local succession of rock groups. Many of these groups yield fossils. If the same fossils are found in different groups in separate areas, it is reasonable to conclude that these particular groups were laid down during the same period of time. Any detailed treatment of fossils is outside the scope of this book, but it is common knowledge that there have been great changes during the geological past in the forms of life which inhabited the earth. By using fossils as time-indices, the geologists have divided the great pile of sedimentary rocks into units called Series and Systems and below that smaller units with which we are not concerned. The term Series is applied to the rocks laid down during an Era, that is one of the major divisions of geological time. Eras are subdivided into Periods, during each of which rocks of a particular System were deposited.

In any one part of the world, one does not find rocks of every system. This could only be the case, if there had been continuous deposition and then the beds had been gently tilted, so that their eroded edges are exposed. As described on p. 84, the stratified rocks are often greatly folded, especially in the mountain chains. The formation of a mountain chain involves great orogenic movements, followed by uplift and erosion of the folded rocks. Therefore on the margins of mountain chains, one finds rocks laid down after the folding movements resting on strongly deformed older strata. The younger beds are said to overlie the older beds unconformably (p. 161). If one knows, from their fossil content, the age of the two groups of rocks, one can determine the age of the period of orogeny, for it must postdate the older group and predate the younger.

Just as the systems have been mostly named after the areas from which they were first described, one can speak of Caledonian, Alpine, etc. Orogenies. Whilst the order and succession of the systems is world wide, orogenic episodes are localized, both in time

89

and space. The number and position in the geological time-scale of orogenic episodes recognized in Europe is different from that demonstrable in America.

During the post-war period, owing to the development of techniques for using radio-active minerals as clocks (below), the geologist has been able to give absolute ages, in millions of years, for the duration of his time divisions (eras and periods). Previous to this development, whilst it was very clear from numerous lines of evidence that the time required for the deposition of a major rock group, such as the Chalk of England, northern France and Germany, must have been very considerable, any estimate of its duration in millions of years, was very much an approximation.

In Table I is set out the geologist's

TABLE I

The Geological Time Scale

ERA	PERIOD	BEGINNING OF PERIOD IN MILLIONS OF YEARS AGO
Quaternary	Pleistocene	2
Tertiary or Cenozoic	Pliocene	7
	Miocene	26
	Alpine Orogeny	
	Oligocene	37
	Eocene	65
Mesozoic	Cretaceous	136
	Jurassic	190
	Triassic	225
Palaeozoic	Permian	280
	Variscan (Armorican) Orogeny	
	Carboniferous	34?
	Devonian	395
	Caledonian Orogeny	
	Silurian	435
	Ordovician	500
	Cambrian	570
Pre-Cambrian	Younger Pre-Cambrian or Algonkian	
	Older Pre-Cambrian or Archaean	
	Origin of Earth	4,700

time-scale or Stratigraphical Table. The major orogenic episodes recognized in Europe have been inserted. The immense duration of the Pre-Cambrian era or eras will be noticed. Owing to the absence of fossils and to the highly metamorphosed nature of the older Pre-Cambrian rocks, it has, as yet, proved impossible to subdivide these beds in anything approaching the detail of the Cambrian and later strata.

The Origin of Mineral Deposits

Mineral deposits are accumulations of one or more substances that are normally present only in very small quantities in the rocks of the earth's crust. These concentrations are often of economic importance. Mineral deposits can occur in a wide variety of forms. Infilling fissures, as roughly cylindrical pipe-like bodies, as definite beds, or as bodies of the most irregular shape. In the metalliferous deposits, the *ore* minerals are mixed with non-metallic minerals, which are known as *gangue*. Usually the ore body is worked for the metallic minerals, such as iron pyrites, copper pyrites, etc., whilst the gangue minerals (quartz, calcite, etc.) are waste products, but this is not always the case. In the mining field of the northern Pennines, it is the gangue minerals, particularly barytes and fluorite, which are profitable today and not the poor quality (lean) lead-zinc ores.

Mineral deposits are of limited extent and are restricted to certain parts of the land areas. Moreover in each mineralised area, one usually finds a particular assemblage of minerals, forming a metallogenetic province. These metallogenetic provinces are generally believed to have originated from differences in the composition of the upper parts of the earth's mantle. Deep crustal flaws have

enabled metal-bearing gases and liquids to rise from great depths into the rocks of the crust.

During the consolidation of magma, concentrations of certain minerals may occur to form *Segregations* interbedded with deep-seated igneous rocks. Examples are the chromite-rich layers of the Bushveld Complex of South Africa and the magnetite deposits of Kiruna, Sweden. In other cases, as in the platinum deposits of the Ural Mountains and in the diamond pipes of South Africa, the economically valuable crystals are scattered *(disseminated)* through the igneous rocks.

As magma crystallizes, the more basic minerals form first, leaving a liquid which becomes increasingly acidic and enriched with the compounds of metals and other valuable substances. This liquid, which is highly charged with volatiles and gases, collects in the upper part of the magma chamber. This residual liquid may react with the 'country rock' to form *contact-metasomatic* deposits, usually very irregular in shape. The gangue-minerals found (garnet, actinolite, epidote, etc.) are high-temperature minerals. The country rock has not been merely baked as in contact metamorphism, but has been considerably changed in chemical composition owing to the addition of material from the magma. Impure calcareous rocks are particularly susceptible to contact-metasomatism.

During the last stages of the consolidation of the magma, highly siliceous molten liquids, may penetrate into the country rocks and the already consolidated igneous rocks to form *pegmatite* veins. Any mineral compounds present in these liquids will crystallize out to form large crystals (See No. 214).

During the final consolidation of the magma, the residual solutions and

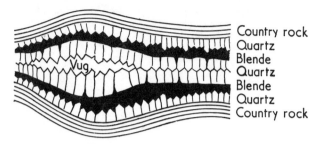

Country rock
Quartz
Blende
Quartz
Blende
Quartz
Country rock

Fig. 7. *A zoned mineral vein with a vug where the vein swells.*

gases will be forced into the cracks and fissures of the surrounding rocks. These *hydrothermal solutions,* at temperatures between 500°C to 50°C, will crystallize wherever circumstances permit. Fissures will be infilled to form *veins* or *lodes,* with the ore and gangue minerals usually showing a banding parallel to the walls of the lodes. Normally the crystals growing at right angles to the walls interfere with one another, but where the fissure was not completely filled, *cavities* or *vugs* occured and these are often lined with well formed crystals (See No. 71 and Fig. 7). The majority of the most beautiful mineral specimens have been obtained from such vugs. The veins vary considerably in width, pinching and swelling, whilst the ore minerals often occur in pockets, separated by lengths of almost barren material. The mineralized veins of any one area normally show a definite pattern, which reflects the stresses to which the host-rocks have been subjected. The majority of veins are relatively short, being traceable for only a few hundred yards, but occasionally they have been followed for miles; for example, one vein has been traced for 10 miles in the Harz

Mountains of Germany, whilst some barren quartz veins in California are believed to be 50 miles in length. Many veins pinch out within a few hundred feet from the surface, but the gold-bearing Champion Lode of Kolar, Mysore, has been mined to a depth of 8,500 feet.

When the country rock is traversed by large numbers of closely-spaced and intersecting veinlets, each one only an inch or two in width and of very limited length, a *stockwork* is produced. (Fig. 8).

Saddle reefs, such as the gold-bearing reefs of Bendigo, Australia, are formed when the ore has been deposited in the cavities along the axes of steeply folded anticlines. A similar structure is produced when a pack of cards is compressed between the hands; the cards do not remain in contact along the crest of the arch that is formed.

Sometimes, in gently dipping or horizontal rocks, the rise of the mineralizing solution has been prevented by some particularly impervious bed. A 'flat' will then be produced, for the solutions spread along the base of the bed. The zones of brecciation along fault planes are often followed by the mineralizing solutions. Angular fragments of rock are cemented together by ore or gangue minerals. (See No. 245).

The mineralizing solutions, instead of infilling fissures and cavities, may remove and then replace the rocks with which they come in contact. Sometimes the *replacement* is complete, elsewhere it is selective and only effects certain constituents of the host-rock or it may be in the form of a dissemination with the ore minerals

Vein

Contact-metasomatic

Breccia filled pipe

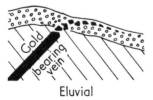

Eluvial

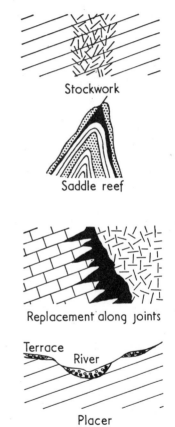

Stockwork

Saddle reef

Replacement along joints

Placer

Fig. 8. *Sections across different types of Mineral Deposits. Mineral deposits, in black; Igneous rock, short dashes and V s; Limestone, bricks; Sand and gravel, circles and dots; Shales and slates, parallel lines.*

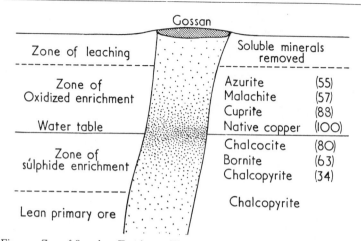

Fig. 9. *Zone of Secondary Enrichment. The intensity of stippling indicates the variation in the copper content of the vein. The figures after the minerals present in each zone are the percentage of copper in pure specimens of each mineral.*

peppered through the host-rock. Some replacement deposits are of enormous size, for example the Rio Tinto deposit of Spain is estimated to contain at least 500 million tons of pyrite, whilst the estimated reserves of the disseminated replacement deposits of the Zambian Copper Belt are even greater.

Secondary Enrichment

Ore deposits, like other rocks, are subject to chemical weathering at their outcrop. In the case of copper lodes, bearing chalcopyrite, their copper content will pass into solution in the zone of leaching and will be carried downwards, leaving behind a capping or *gossan,* to use the Cornish miners' term, made of insoluble material, mainly iron compounds. Many mineral veins, rich at depth, have been found by prospectors, searching for the

distinctive gossan. The copper carried downwards may be deposited in the zone of secondary enrichment, which can be divided into an upper zone, above the water table, where carbonates, oxides and the pure mineral are deposited, and a lower zone of sulphide enrichment. Below this enriched zone, the ore reverts to its original primary state (Fig. 9) and may be too lean to repay working. Silver ores are enriched in the same way, but the other metallic sulphides pass into solution much less easily.

Placer Deposits

Weathering, whether mechanical or chemical, loosens minerals from their matrix. During the process of the downhill movement of the concentrated rock material, particularly heavy and stable minerals, such as native gold, native silver, native platinum and cassiterite (tinstone) may become concentrated. Moving water sweeps away the lighter material, so that eventually the sparse gold content of thousands of tons of rock material may be concentrated in the bottom of a pool in a river. Placer deposits may be *eluvial*, formed without stream action, on the downslope side of rich lodes, or they may be *alluvial*, in the sands and gravels laid down by streams, (Fig. 8) or they may be *beach* placers, where the sorting has been effected by waves. Many of the great 'gold rushes' such as the California gold rush of 1849 and the later stampede to the Klondyke, were due to the discovery of rich alluvial deposits. In 1926 the discovery of diamond-bearing gravels caused a great rush to Lichtenburg in South West Africa. Gemstones (rubies, sapphires, etc.) have been concentrated in stream gravels in Ceylon, Kashmir,

North Carolina, etc. Beach sands may also contain unusual concentrations of certain normally rare minerals. For example, at Travancore, India, the beach sands contain up to 25 % of monazite, a cerium and thorium-rich mineral. Black beach sands may be due to concentrations of magnetite, ilmenite, chromite, etc.

Mineral Deposits of Sedimentary or Metamorphic Origin

The sedimentary rocks of economic importance are described in the latter part of the book. As a result of either contact or regional metamorphism deposits rich in such metamorphic minerals as asbestos, talc, graphite, garnet, andalusite, kyanite, etc., may be formed. These will be described under the heading of the appropriate mineral.

PLATE TECTONICS

The development of techniques for drilling into and obtaining samples of the rocks of the ocean floors, the improvements in rock dating and of palaeomagnetism (determining the position of a rock relative to the earth's magnetic field at its time of formation) have produced a revolution in geological thought. It is now generally recognized that the earth's crust is composed of a number of plates, some of oceanic crust, others largely of continental crust, the remainder of both oceanic and continental crust. Throughout the later part, at least, of geological time, these plates have both changed their shape and also have been in movement relative to one another.

The Atlantic Ocean has developed during the past 200 million years by the drifting apart of the American and

of the European and African plates. They have parted along the Mid-Atlantic Ridge, a deeply submerged rift valley. Magma has well up along this line of *Sea Floor Spreading*, which has been opening at a rate measurable in centimetres a year. The active volcanoes of Iceland and the volcanic islands of the Azores, Ascension and Tristan da Cunha are situated along this ridge.

Where two plates move towards each other, the rocks in any geosyncline marginal to one of the plates, will be contorted and metamorphosed (p. 89). The Alps were formed by the northward drive of the African and minor plates, the much older Caledonian chains by the closing of the plates on either side of a Proto-Atlantic.

95

II. THE IDENTIFICATION
OF MINERALS

Minerals are identified, as are most other objects, man-made or natural, by the presence of a particular combination of characters. It is, therefore, necessary to describe first the variation in each of the characters of minerals and the descriptive terms used. Then for each mineral, we must select those combinations that are diagnostic of that particular mineral. It must, however, be emphasized that most minerals may show quite a variation in certain characters. The different varieties of quartz, for example, cover a wide range of colour (p. 123), but they all have other properties in common. Again many minerals may occur in quite a variety of shapes, but whatever their shape, specimens of each mineral will have other properties in common.

In Table II are given the chief diagnostic characters of the commoner varieties of the minerals described and illustrated in this book. It must, however, be emphasized that these tables should not be regarded as comprehensive. One is always liable to find an unusual variety of a common mineral. Should this occur comparison of the table with the colour plates will give a good indication of what the mineral may be. This can be confirmed by consulting an expert or by visiting a museum with a comprehensive collection of minerals on display. Poorly preserved specimens of minerals may well be difficult to identify.

Crystal Form

Under conditions favourable to crystal growth, minerals will occur in the *crystallized* form with well developed crystals (e.g. No. 9). If, however, the growing crystals interfere with

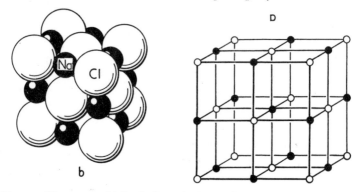

Fig. 10. *The structure of halite (sodium chloride): a. The crystal lattice with position of sodium (Na) ions in black and chlorine (Cl) in white, b. Drawn to scale of* 40,000,000:1.

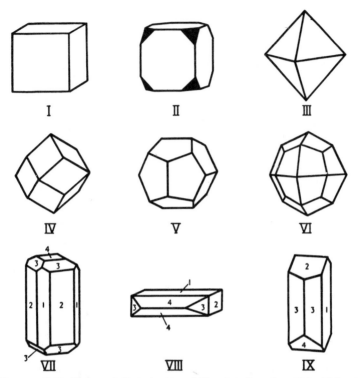

Fig. 11. *Cubic, Tetragonal, Orthorhombic and Monoclinic Crystals.* I *Cube (6 faces), e.g. pyrites, halite, galena, fluorite,* II *Cube with edges bevelled by faces of the octahedron, e.g. fluorite,* III *Octahedron (8 faces) e.g. diamond,* IV *Rhombic dodecahedron (12 faces) e.g. garnet,* V *Pentagonal dodecahedron or pyritohedron (12 faces) e.g. pyrites,* VI *Icositetrahedron (24 faces) e.g. pyrites,* VII *Tetragonal crystal of idocrase with faces of two prisms (1 and 2), pyramid (3) and basal pinacoid (4),* VIII *Orthorhombic crystal of barytes with faces of basal and vertical pinacoids (1 and 2) and of vertical and horizontal prisms (3 and 4),* IX *Monoclinic crystal of orthoclase feldspar with faces of side and basal pinacoids (1 and 2), prism (3) and hemidome (4).*

each other, then they may be distorted or malformed. The mineral will then be in the *crystalline* form (e.g. No. 12), for even if the crystals are imperfectly formed, the mineral has an ordered

atomic structure and a definite chemical composition.

Crystals are geometric forms with flat faces arranged on a definite plan, which is an expression of the atomic

97

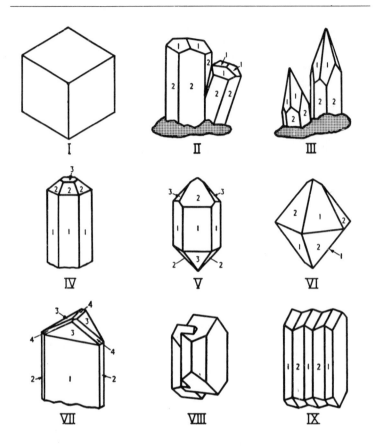

Fig. 12. *Hexagonal, Trigonal and Twinned Crystals.* i *Rhombohedron (6 faces), e.g. dolomite,* ii *Nail-head calcite with rhombohedra (1) capping hexagonal prisms (2),* iii *Dog-tooth calcite with scalenohedra (1) capping hexagonal prisms (2),* iv *Hexagonal prism (1) with hexagonal pyramid (2) terminated by basal pinacoid (3), e.g. apatite,* v *Hexagonal prism (1) terminated by two unequally developed rhombohedra (2 and 3), e.g. quartz,* vi *Two equally developed rhombohedra (1 and 2), e.g. quartz,* vii *Trigonal crystal made up of two trigonal prisms (1 and 2) and two trigonal pyramids (3 and 4) e.g. tourmaline,* viii *Carlsbad twin of orthoclase feldspar, cf. Fig.* 11 viii, ix *Repeated twinning of plagioclase feldspar.*

structure of the mineral (Fig. 10). In Crystallography, the study of crystals, the faces of a crystal are related to a number of imaginary crystal axes meeting at a point in the centre of the crystal. As shown below, the assumed relative length and inclination of these axis is different in each of the seven Crystal Systems.

Detailed crystallography is beyond the scope of this book, but in Figs. 11 and 12, are shown a selection of the commoner forms of crystals to indicate the chief differences between those crystallizing in each system. Each mineral crystallizes in a particular system and often only occurs in a limited number of forms, e.g. diamond, when crystallized, commonly occurs as

octahedra, though two octahedra may be joined together to form a twin crystal.

A *twin crystal* is made up of two or more crystals of the same mineral having a different orientation. The halves of the twin are related to one another according to a definite law. Twins may be interpenetrant (as in orthoclase feldspar, Fig. 12 VIII or staurolite, Fig. 26), rotation twins (as in gypsum, Fig. 22), they may show repeated twinning (as in plagioclase feldspar, Fig. 12, IX) or be geniculate (knee shaped) (as in rutile, Fig. 17). The presence of reentrant angles between certain of the crystal faces is a clear indication of twinning.

The relation of the axes, some of the commoner forms and the typical minerals crystallizing in each of the crystal systems are shown below:

SYSTEM	AXES		TYPICAL FORMS	MINERALS
Cubic	3 equal axes, all at right angles		Cube, octahedron, dodecahedron, tetrahedron	diamond, galena, pyrite, garnet, galena, blende.
Tetragonal	3 axes at right angles, the two horizontal axes of equal length, the vertical one of different length		4 or 8 faced prisma and pyramids, 2 faced basal pinacoids	chalcopyrite, rutile, zircon, cassiterite
Hexagonal	1 vertical and 3 horizontal axes at angels of 120° to each other. The horizontal axes of equal length.		6 or 12 faced prisma and pyramids, 2 faced basal pinacoids.	pyrrhotite, apatite, beryl

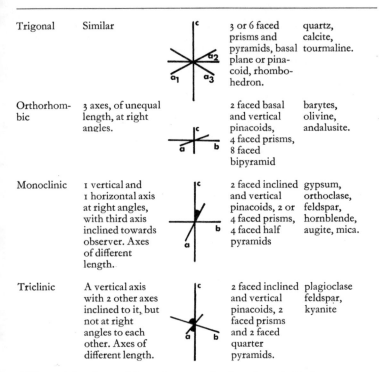

Trigonal	Similar		3 or 6 faced prisms and pyramids, basal plane or pinacoid, rhombo- hedron.	quartz, calcite, tourmaline.
Orthorhom- bic	3 axes, of unequal length, at right angles.		2 faced basal and vertical pinacoids, 4 faced prisms, 8 faced bipyramid	barytes, olivine, andalusite.
Monoclinic	1 vertical and 1 horizontal axis at right angles, with third axis inclined towards observer. Axes of different length.		2 faced inclined and vertical pinacoids, 2 or 4 faced prisms, 4 faced half pyramids	gypsum, orthoclase, feldspar, hornblende, augite, mica.
Triclinic	A vertical axis with 2 other axes inclined to it, but not at right angles to each other. Axes of different length.		2 faced inclined and vertical pinacoids, 2 faced prisms and 2 faced quarter pyramids.	plagioclase feldspar, kyanite

All interaxial angles are right angles, except for those shown in black.

It will be noticed that the forms of the Cubic System are closed forms. In the other systems it is necessary to combine two or more forms, e.g. Tetragonal or Hexagonal prism and basal pinacoid to produce a solid crystal, whilst in the Monoclinic and Triclinic Systems of still lower symmetry even more forms may have to be combined (e.g. orthoclase feldspar, Fig. 11, IX).

The most regular crystals are to be found in the Cubic System, there the atoms are the most tightly packed. Then as one comes down the scale, their regularity or degree of symmetry, is reduced until one reaches the Triclinic System where the crystals are the least regular. Crystals are built of atoms arranged on a pattern known as 'the space lattice'. In exactly the same way as the pattern on wall-paper repeats itself, so it does in any parti- cular mineral. In micas, the majority of the atoms lie in thin layers one above the other. The micas, therefore, can be split, or cleaved, very easily into thin sheets. In actinolite, the atoms lie in long chains, very loosely linked together. Thus actinolite, and in particular asbestiform actinolite, is easily split into long needles or fibre. In diamonds and garnets the atoms lie tightly packed in all directions. For that reason they form hard cubic crystals.

Form or Habit

Minerals also occur in a variety of forms, which are not directly dependent on their crystal structure. The names of certain forms, e.g. columnar, nodular, radiating and fibrous are self explanatory, other terms are briefly defined below:-

amygdaloidal – almond-shaped, infilling steam-cavities in the top of lava flows. (No. 186).

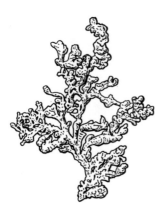

Fig. 13. *Arborescent copper.*

arborescent – moss-like markings on the walls of a narrow crack. Shown by native copper (Fig. 13).

botryoidal – like a bunch of grapes.

dendritic – tree-like markings on a crack or bedding plane shown by pyrolusite (No. 274). A common habit of chalcedony.

filiform – thin wires, often twisted together like the strands of a rope, often shown by native copper.

foliated – in thin, easily separable leaves, as in mica.

granular – in grains. If the grains are of equal size, the term saccharoidal is used, because of the resemblance to lump sugar.

mammillated – large sphaeroids, which intersect each other, as in malachite. (No. 75).

reniform – with rounded kidney-shaped surfaces, as in kidney iron ore. (Fig. 18).

tabular – with broad flat surfaces.

Other Physical Properties

Colour is only diagnostic in a few special cases, e.g. amethystine quartz. Many minerals show a considerable range in colour. Colour can, however, often be of considerable value, when combined with some other property. There are many white minerals, but only barytes or heavy spar, feels significantly heavy in the hand.

The **Hardness** of a mineral is measured in terms of Mohs' Scale, an arbitrary scale, with very irregular intervals.
Hardness (H)

1. Talc (can be crushed by finger nail)
2. Gypsum (scratches with finger nail)
3. Calcite (scratches with iron nail)
4. Fluorite (scratches with glass)

5. Apatite (scratches with penknife - ordinary steel)
6. Orthoclase feldspar (scratches with quartz)
7. Quartz (scratches with file - special steel)
8. Topaz and beryl (scratched by emerald)
9. Corundum (emerald, sapphire, ruby, only scratched by diamond)
10. Diamond

In some instances the hardness may vary slightly on different faces of one and the same crystal (exceptionally this difference may be considerable, in kyanite from 4 to 7).

The weight of a mineral is stated in terms of its **Specific Gravity** (S.G.) that is its weight as compared to that of the same volume of water. Minerals whose S.G. is much above 3 feel noticeably heavy.

Cleavage is the tendency to split easily along certain planes which are related to the crystalline form and atomic structure of the mineral. Mica for example splits into thin sheets along a highly developed (perfect) cleavage parallel to the base of the crystals. Fluorite crystallizes in cubes, but the corners of the cubes are often bevelled, owing to the mineral cleaving along faces parallel to those of an octahedron. (Fig. 11, II).

Some minerals such as talc have a distinctive soapy feel, others such as rock salt a characteristic taste, others

may show a particular **lustre**, such as a diamond. The terms commonly used for describing *lustre* are set out below:-
Metallic
Vitreous (of broken glass)
Resinous
Pearly
Silky
Adamantine (of a diamond)
In addition the adjective dull, splendent, etc., can be used.

The **Streak** of some minerals, particularly the iron oxides and hydroxides, is distinctive. A soft mineral will produce its streak, when rubbed on unglazed porcelain or stout paper. With a harder mineral, it is usually necessary to scratch the mineral with a knife, to produce a powder and then to crush that.

The colour plates 1-158 illustrate typical examples of the commoner minerals, together with a limited number of polished gemstones. The minerals have been arranged with the pure elements (e.g. the native metals) first followed by combinations of element and sulphur (sulphides) and element and oxygen (oxides), very often with water as a constituent. (hydroxides or 'hydrates'). Then follow the salts, – chlorides (element and chlorine or fluorine), the carbonates (element, carbon and oxygen), the sulphates (element, wolframites and silicates (element, silicon and oxygen) in that order. Last come the radioactive minerals.

TABLE II
THE IDENTIFICATION
OF MINERALS

A – Accessory Minerals of Igneous
Rocks
E – Essential Minerals of Igneous
Rocks
G – Common Gangue Minerals
M – Minerals indicative of
Metamorphism
O – Common Ore Minerals
P – Precious Stones
S – Semi-precious Stones

NAME	COMPOSITION	CAT-EGORY	COLOUR	COMMON FORM	OTHER DISTINCTIVE PROPERTIES	FIG.	PAGE
	Native Elements						
Gold	Au		Golden yellow, does not tarnish	Commonly as scales or water-worn grains and nuggets. Cubic crystals rare	S.G. 15-20, H. 2.5-3	1 & 2	117
Silver	Ag		Silver-white, tarnishes brown	Massive or arborescent. If crystalline distorted cubes or octahedra	S.G. 10-12, H. 2.5-3	3	117
Copper	Cu		Copper-red, but darkens on exposure	Massive, arborescent or as threads	S.G. 8.8, H. 2.5-3	4	118
Iron	Fe		Iron-grey	Usually massive, if crystalline as octahedra. Found in meteorites	S.G. 7.5, strongly magnetic		118
Diamond	C	P	White or colourless, industrial diamonds black or other colours	Octahedral crystals, waterworn in placer deposits	H. 10, adamantine lustre	5 & 6	118
Graphite (Black lead)	C		Iron-grey, black	Scales, sometimes granular	H. 1-2, soils fingers, S.G. 2.2-2.3, metallic lustre	7	119

Sulphur	S		Sulphur-yellow	Orthorhombic crystals, also massive and encrusting	Resinous lustre, H. 1.5-2.5, S.G. 2	268	119
Sulphides							
Pyrites	FeS_2	O	Bronze-yellow, does not tarnish	Cubic crystals, often as striated cubes, or as pyritohedra, also massive or nodular	Splendent metallic lustre, H. 6-6.5	8-12	120
Marcasite	FeS_2		Paler than pyrites	Tabular orthorhombic crystals, often twinned or as nodules with radiating structure	Metallic lustre, H. 6-6.5	272	120
Pyrrhotite (Magnetic pyrites)	FeS with some nickel	O	Reddish, but tarnishes easily	Usually massive, rarely as hexagonal prisms	H. 3.5-4.5, magnetic	13	120
Mispickel (Arsenical pyrites)	FeAsS	O	Steel-grey	Striated orthorhombic crystals, also massive	Metallic lustre	14	121
Realgar	AsS	O	Red to orange	Usually massive. If crystalline, prismatic monoclinic crystals	Resinous lustre, H. 1.5-2	269	121
Orpiment	As_2S_3	O	Lemon-yellow	Usually massive or foliated. The crystals are monoclinic	Pearly lustre on crystalline faces, otherwise resinous lustre	270	121

NAME	COMPOSITION	CAT-EGORIE	COLOUR	COMMON FORM	OTHER DISTINCTIVE PROPERTIES	FIG.	PAGE
Chalcopyrite (Copper pyrites)	Cu_2S. Fe_2S_3	O	Brass-yellow, but often tarnished, sometimes iridescent	Tetragonal crystals, also massive	Compare colour with pyrite, H. 3.5-4	16	121
Bornite (Variegated copper ore)	Sulphide of Fe and Cu	O	Copper-red, tarnishes easily to Peacock ore	Usually massive, cubic crystals rare	H. 3	17	121
Chalcocite (Copper glance)	Cu_2S	O	Lead-grey, often with bluish tinge	Usually massive, orthorhombic crystals rare	H. 2.5-3	18	121
Galena (Lead glance)	PbS	O	Lead-grey	Cubic crystals, perfect cubic cleavage. Also massive and granular	Dull metallic lustre, S.G. 7.5, H. 2.5	19-20	122
Cinnabar	HgS	O	Cochineal-red	Usually massive. Hexagonal crystals rare	S.G. 8 Adamantine lustre on crystals, but massive forms dull	271	122
Blende (Sphalerite, black jack)	ZnS	O	Brown, black or red	Usually massive. Tetrahedral cubic crystals, often twinned	Adamantine lustre on crystals, resinous lustre on massive forms, H. 4	21-22	122

Name	Formula		Colour	Form / Habit	Lustre / Properties	No.	Page
Molybdenite	MoS_2		Lead-grey	Usually as scales, also massive	H. 1-1.5 S.G. 4.7	23-24	122
Stibnite (Antimony glance)	Sb_2S_3		Lead-grey	Elongated and striated orthorhombic crystals or as radiating masses	H. 2. Metallic lustre but tarnishes easily and then becomes iridescent	25	123
Cobaltite	$CoAsS$		Silver-white	Cubic crystals or massive	Metallic lustre, H. 5.5, S.G. 6	26	123
Oxides							
Quartz	SiO_2	E G S	When pure colourless, but many coloured varieties (p. 116)	Crystals with hexagonal prisms and rhombohedra, also massive	Vitreous lustre, H. 7	27-35	123
Chalcedony	Mixture of quartz and opal		Varied (see p. 115)	Usually infills cavities or as nodules with botryoidal surface	Waxy lustre, cryptocrystalline	36-40	124
Opal	$SiO_2 \cdot nH_2O$	P	Varied colours showing beautiful opalescence	Amorphous	Subvitreous lustre, H. 5.5	41	124
Rutile	TiO_2	A O	Reddish-brown or black	Tetragonal crystals, often striated, twins common	Metallic lustre	45	125

NAME	COMPOSITION	CATEGORY	COLOUR	COMMON FORM	OTHER DISTINCTIVE PROPERTIES	FIG.	PAGE
Cassiterite (Tinstone)	SnO_2	O	Black	Tetragonal crystals, also massive. Water worn grains (Stream tin)	Adamantine lustre, S.G. 7, H. 6-7	46	125
Cuprite	Cu_2O	O	Shades of red	Octahedral and rhomb-dodecahedral crystals, also massive	Crystals show adamantine lustre, massive forms submetallic. S.G. 6, streak brownish-red		125
Corundum	Al_2O_3	P	Common varieties grey and dull, but for gem varieties see p. 118	Trigonal crystals, often water worn, also occurs massive	H. 9	47-49	126
Spinel	$MgO.$ Al_2O_3	S M	Red, brown or black	Octahedral crystals	H. 8	277	126
Magnetite (Magnetic iron ore)	Fe_3O_4	A O	Iron-black	Octahedral crystals, also massive	H. 6, magnetic	52-56	125
Hematite (Specular iron ore, Kidney ore)	Fe_2O_3	O	Black in Specular iron, red in Kidney ore, but surface often tarnishes to black	Rhombohedral crystals in Specular iron. Kidney ore reniform and fibrous	Lustre of Specular ore metallic and highly splendent, silky lustre in fibrous varieties, streak-cherry-red	50-51 273	125

Name	Formula		Colour	Form / Habit	Lustre / Hardness		
Limonite (Brown hematite)	$2Fe_2O_3 \cdot nH_2O$	O	Shades of brown	Mammillated or stalactitic with fibrous radiating structure, also earthy or oolitic	Lustre submetallic or dull, streak yellow-brown	58	127
Ilmenite	$FeO.TiO_2$	O A	Iron-black	Usually massive or in scales, if crystalline as complex trigonal rhombohedra	Submetallic lustre	58	127
Braunite	Mn_2O_3 with some silica	O	Brownish-black	Tetragonal crystals or massive	Submetallic lustre	57	127
Manganite	$Mn_2O_3.H_2O$	O	Iron-black	Striated orthorhombic prismatic crystals	Submetallic lustre	62	127
Pyrolusite	MnO_2		Iron-grey	Usually massive, sometimes dendritic	H. 2	274-75	127
Halides							
Rock Salt (Halite)	NaCl		Colourless if pure usually bluish	Cubes, sometimes with hollow faces, also massive or granular	Taste, H. 2	63	128
Fluorite (Fluor Spar)	CaF_2	G	Variable, Blue John is the purple variety	Cubes with octahedral cleavage. Also massive	H. 4. Vitreous lustre	64-65	129
Cryolite	$AlF_3.NaF$		Variable	Monoclinic crystals which appear to be cubic, but usually massive.	Almost invisible in water. H. 2.5	66	129

NAME	COMPOSITION	CAT-EGORY	COLOUR	COMMON FORM	OTHER DISTINCTIVE PROPERTIES	FIG.	PAGE
Apatite	Phosphate and fluoride or chloride of calcium	A	Variable, but usually pale green	Hexagonal crystals, also mammillated and massive	Subresinous lustre H. 3	82-83	130
Carbonates							
Calcite (Calc Spar)	$CaCO_3$	G	Usually white	Hexagonal crystals, good rhombohedral cleavage. Also in other habits (p. 122)	Effervesces with cold dilute hydrochloric acid	67-72	130
Aragonite	$CaCO_3$		White or greyish	Orthorhombic crystals, often with reentrant angles owing to twinning, also massive or fibrous	H. 3.5-4	73	130
Dolomite (Pearl Spar)	$CaCO_3 \cdot MgCO_3$	G	White, often tinged with yellow	Hexagonal crystals usually rhombohedra with curved faces. Also massive and granular. Perfect rhombohedral cleavage	Pearly lustre on crystal faces, H. 3.5-4, does not effervesce with cold dilute hydrochloric acid, compare calcite	278	131

Magnesite	$MgCO_3$		White	Usually massive or fibrous	Earthy lustre if fibrous, otherwise vitreous		131
Siderite (Chalybite)	$FeCO_3$		Shades of brown	Curved rhombohedra, perfect rhombohedral cleavage. Also massive	Pearly or vitreous lustre	74	131
Malachite	$CuCO_3 \cdot Cu(OH)_2$	O	Bright green	Surface usually mammillated, fibrous internally	Banded with silky lustre on broken edges	75	131
Azurite	$2CuCO_3 \cdot Cu(OH)_2$	O	Azure blue	Monoclinic if crystalline, but usually massive	Colour very distinctive	76	131
Witherite	$BaCO_3$	O	White or yellowish	Twinned orthorhombic crystals, also massive	Lustre vitreous to dull		131
Sulphates							
Barytes (Heavy Spar)	$BaSO_4$	G	White, but often tinged with yellow or brown	Flat orthorhombic crystals with perfect basal and prismatic cleavage, also massive	Heavy (S.G. 4.5), vitreous to resinous lustre	77	132
Gypsum	$CaSO_4 \cdot 2H_2O$		White	Monoclinic crystals (Selenite), compact form (Alabaster), fibrous form (Satin Spar)	H. 2, pearly lustre on crystal faces, fibrous varieties have silky lustre	78-80	132

111

NAME	COMPOSITION	CATEGORY	COLOUR	COMMON FORM	OTHER DISTINCTIVE PROPERTIES	FIG.	PAGE
Anhydrite	$CaSO_4$		White, often tinged with grey or blue	Orthorhombic crystals, also compact, fibrous and granular	H. 3, pearly lustre on cleavage planes		132
	Wolframites						
Scheelite	$CaWO_4$	O	Shades of yellow	Tetragonal crystals, also reniform and massive	Vitreous lustre, S.G. 6	81	133
	Silicates						
Orthoclase Feldspar	$KAlSi_3O_8$	E	White or flesh coloured	Monoclinic crystals, simple twinning, good cleavage, also massive	H. 6, vitreous to pearly lustre	84, 85 87	133
Amazonstone	$KAlSi_3O_8$		The green variety of microcline	Triclinic crystals, good cleavage, also massive	Distinctive colour, lamellar twinning	86	134
Plagioclase Feldspar	Mixtures of Ca and Na, Al silicates	E	Usually white	Triclinic crystals, repeated twinning, also massive	Repeated twinning distinctive	88, 89	134
Leucite	$KAlSi_2O_6$		White or ash grey	Cubic crystals, also in disseminated grains	8-sided cross-section	90	135
Lapis Lazuli	Na, Al silicate with sodium sulphide	S	Blue	Usually massive, cubic crystals rare	Distinctive colour	276	135

Nepheline	K, Na, Al silicate		White, grey or colourless	Six-sided hexagonal crystals or massive	Vitreous or greasy lustre	91	135
Scapolite	Al, Na, Ca silicate with NaCl and $CaCO_3$		White, often faintly tinged	Tetragonal crystals or massive	Vitreous to pearly lustre	92	138
Garnet	Al silicates with Ca, Mg, Mn or Fe	M	Lime Garnet green, other garnets shades of red	Cubic crystals, often dodecahedra and icositetrahedra	H. 7	93-96 279	138
Olivine	Mg, Fe silicate	E	Shades of green, yellow or brown	Orthorhombic crystals, also massive or granular triclinic	H. 7	97	139
Rhodonite (Manganese Spar)	$MnSiO_3$ with some Fe, Ca, Mg or Zn	M	Red, but darkens on exposure	Tabular triclinic crystals but usually massive. Good cleavage	Vitreous lustre	111	139
Vesuvianite (Idocrase)	Ca, Al silicate with some Fe, Mg, etc.	M	Deep browns and greens	Well formed prismatic tetragonal crystals, also massive		98	139
Zircon	$ZrSiO_4$	A S	Colourless, yellow, green or red	Tetragonal crystals, also as in waterworn grains	Adamantine lustre. H. 7.5	99	139
Topaz	$Al_2F_2SiO_4$	M S	Yellow, blue, green or colourless	Prismatic orthorhombic crystals, also granular	H. 8	100-101	140
Andalusite	Al_2SiO_5	M	Red, but often surface alteration into silvery mica	Orthorhombic prismatic crystals, also massive	H. 7.5, variety chiastolite as cruciform crystals with dark centres	102	140

NAME	COMPOSITION	CAT-EGORY	COLOUR	COMMON FORM	OTHER DISTINCTIVE PROPERTIES	FIG.	PAGE
Sillimanite	Al_2SiO_5	M	Brown or grey	Long needle-shaped orthorhombic crystals		103	141
Kyanite (Disthene)	Al_2SiO_5	M	Light blue	Long bladed triclinic crystals, also radiating rosettes	Pearly lustre. H varies from 4-7 on different faces	104	141
Epidote	Ca, Al, Fe silicate	M S	Shades of green	Elongated monoclinic crystals	Perfect basal cleavage	105-106	141
Pyroxene Family	Silicates of Fe, Mg, and Ca, sometimes with Al	E	Usually shades of dark green to black	Stout orthorhombic crystals (Enstatite). Prismatic monoclinic crystals (Augite), also massive	H. 5-6	107-110	135
Amphibole Family	Silicates of Ca, Mg, and Fe, sometimes with Al	E	Usually brown to black, fibrous varieties green or blue	Prismatic monoclinic crystals, also massive. Asbestiform varieties fibrous		38, 112-118	136
Beryl	Silicate of beryllium and Al	M S P	Green (emerald), pale blue (aquamarine), yellow and white	Prismatic hexagonal crystals	H. 7.5 Vitreous to resinous lustre	120-124	141
Tourmaline	Silicate of boron and Al with alkalies	M	Black, blue, green, or red	3-sided striated trigonal crystals, also massive and columnar	Colours often arranged in zones. Ends of crystals may be differently coloured	125-128	141

					Colour		
Cordierite	Silicate of Al, Fe, Mg with water	M	Blue	Short 6-sided crystals also massive		129-130	142
Staurolite	Hydrated Al, Fe silicate	M	Brownish or reddish	Orthorhombic crystals, twins interpenetrant	Crystals usually have dull rough surfaces	131	142
Apophyllite	Hydrous silicate of Ca and K		White or grey	Tetragonal crystals, also massive	A zeolite infilling cavities and amygdales in lavas	132	142
Natrolite	Hydrous silicate of Na and Al		White or greyish	Slender needle-like crystals, also massive or fibrous	In amygdales of basalts	133-4	142
Muscovite (common mica)	Silicate of Al, K and H	M	White	6-sided (pseudo-hexagonal) monoclinic crystals with perfect basal cleavage	Pearly lustre, plates elastic and flexible	135	138
Biotite (dark mica)	Silicate of Mg, Fe, K and H	E M	Black or dark green	Like muscovite	Splendent lustre but pearly on cleavage faces, plates elastic and flexible	138-9	138
Lepidolite (lithia mica)	Silicate of Al, K and lithium		Pink or lilac	Like muscovite	pearly lustre	136-7	138

NAME	COMPOSITION	CATEGORY	COLOUR	COMMON FORM	OTHER DISTINCTIVE PROPERTIES	FIG.	PAGE
Fuchsite (chrome mica)	Silicate of Al K and chromium		Green	Like muscovite	pearly lustre	140-1	138
Chlorite	Hydrated Al, Fe, Mg, silicate	M	Shades of green	Tabular crystals, but usually in scales	Slightly greasy feel, H. 1.5-2.5 Scales flexible but not elastic, compare micas	142	142
Serpentine	Hydrated Mg. silicate	M	Shades of green, brown, black, red or yellow	Massive, granular or fibrous	H. 3-4 Colours show veining or brecciation	143-144	143
Talc	Hydrated Mg. silicate	M	White or green	Usually massive (Soapstone), also granular or in scales	Pearly lustre, greasy feel, H. 1	145	143
Glauconite	Hydrated Fe, Al and K silicate		Green	Granular	Dull lustre easily oxidizes	146	143
Sphene (Titanite)	CaO. TiO$_2$, SiO$_2$	A	Brown or yellow	Wedge-shaped monoclinic crystals, also massive	Adamantine or resinous lustre, H. 5	147	143
Monazite	Complex phosphate of cerium and thorium minerals	A	Yellow or brown	Monoclinic crystals, also massive or as waterworn grains	Resinous lustre	152	144

116

DESCRIPTION OF THE COMMONER MINERALS

NATIVE ELEMENTS

Gold (Au, Nos. 1 and 2, and Table p. 104) is recognizable by its colour (thought if it contains much silver it is much paler), and its weight (S.G. 15-20). The variation in weight depends on the proportion of other metals with which the gold is alloyed.

Gold-bearing (auriferous) quartz veins associated with granite masses have been worked in many parts of the world (India, Canada, Brazil, Ghana, etc.). Small specks of gold are usually visible in the quartz veins, but in some mines the gold, in particles too small to be seen by the unaided eye, is enclosed in sulphide minerals such as iron pyrites, pyrrhotite, and copper pyrites. Gold-bearing placer deposits, sometimes ancient river terraces high on the sides of the present valleys, were discovered in California in 1849, in the Klondyke (Alaska) in 1898 and caused the notorious 'gold-rushes'. Other rich placer deposits occur in Victoria, Australia and in Siberia. At Nome in Alaska, beach placers are worked. In these placer deposits, the gold usually occurs in the lower levels and varies in size from mere flecks to nuggets of many pounds weight. The famous 'Welcome Stranger' nugget discovered in 1869 at Ballarat, Australia, weighed 2280 ounces and was valued at £10,000. It was found in the rut made by a cart.

In the Rand gold-fields of the Transvaal, South Africa, the gold occurs in a Pre-Cambrian conglomerate, the 'banket'. This conglomerate may be a 'fossil' placer deposit, but by some it is believed that the gold was deposited later than the conglomerate by mineralizing solutions, which carried as well as the gold, the other sulphides, such as iron pyrites, pyrrhotite and cobaltite, which are scattered through the 'banket'.

In the United Kingdom there are a few localities from which some gold has been obtained in the past, mainly from quartz veins. One of the Welsh Mines, near Dolgellau in Merionethshire, was specially re-opened to obtain sufficient gold to make the wedding ring for Queen Elizabeth II. This indicates the almost barren nature of the British 'gold-bearing' veins.

Silver (AG, No. 3. and Table p. 104) is recognized by its colour, though it tarnishes readily, and by its weight (S.G. 10-11).

Mineral veins carrying native silver have been mined at many localities in the western mountain chains of North and South America. The richest ore occurs in pockets or 'bonanzas'. One shoot of the Camstock Lode in Nevada produced in three years silver to the value of more than £20 million. In Mexico nuggets of almost pure silver, over one ton in weight, have been found. Much secondary enrichment of leaner primary ore has occurred. In the Lake Superior region of Canada native silver occurs as an infilling of steam vesicles in lava flows.

The bulk of the world's silver production comes, however, not from native silver, but from argentite (silver sulphide) and other silver minerals which are found in association with sulphide ores, especially those of lead, zinc and copper. Argentiferous (silver-bearing) lead-zinc ores are mined at Broken Hill, New South Wales, in British Columbia, Burma, Peru, Utah, Sudbury in Canada, etc. In parts of the British Isles (the Mendips, the Snailbeach area of Shropshire, the northern Pennines, Leadhills in Southern Scotland), argentiferous lead ores have been mined in the past, but the yield of silver was never appreciable.

Lead (Pb) is very rarely found in the native state.

Native Copper (Cu, No. 4. and Table p. 104), on the other hand, is not so infrequent. It occurs either in the massive form or as arborescent sheets (Fig. 13) recognizable by their colour and weight (S.G. 8.8). The richest deposits of pure copper are in the Lake Superior region of Canada, where the native metal occurs in conglomerates, replacing both pebbles and matrix, in the steam cavities of lava flows, and also in small veins. In the other copper mining fields, such as those of the southwestern United States, Chile, Peru, Canada, Rhodesia and Katanga, the mineral is obtained mainly from sulphide and carbonate ores, though native copper may occur in small amounts in the secondarily enriched upper parts of the ore bodies.

Native Iron (Fe) is of extremely rare occurrence in terrestial rocks. It is however, an important constituent of certain meteorites (see p. 173).

Diamond (C, Nos. 5 and 6 and Table p. 104). Pure carbon is found in nature in two forms – hard clear diamond and soft dirty graphite (see below). Diamond is the hardest of all known minerals. Its lustre is unique and its resistance to chemical attack very high. For these reasons diamond has become the most highly prized of gem stones. The biggest known diamond is the Cullinan found in South Africa in 1905. It originally weighed 620 grammes (one and one third pounds). It was cut into several stones, which are now included in the British Crown Jewels. Another British Crown Jewel is the Kohinoor diamond (No. 5), which was discovered a long time ago in India. When it was cut in 1852, its weight was reduced to almost half that of the original 'rough' diamond.

The cutting of diamonds so that they may display their brilliance to the best advantage, is an extremely skilled operation. Two of the many varieties of cut used are shown on Fig. 14. The normal loss during the cutting and polishing of gem quality diamonds is between 50 and 60 %, though it may be more if the stone is badly flawed.

Diamonds of gem quality should be

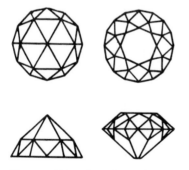

Fig. 14. *Examples of cut diamonds. On left - Dutch rose) On right - Brilliant. In each case, crown view above and side view below.*

as nearly colourless as possible, but coloured stones (No. 6) are quite attractive. Diamond also occurs as granular or rounded aggregates. In this form it is known as Bort or Carbonado, if black. Bort and gem diamond of inferior quality are used extensively for abrasive and cutting purposes, e.g. in diamond drills, diamond wheels, etc.

The majority of gem diamonds come from the Union of South Africa, Angola, and the Congo, and of bort from Brazil. The most famous mines are at Kimberley, South Africa where the stones are embedded in an igneous rock called Kimberlite when it is un-weathered, and Blue Ground (No. 6) when weathered. The kimberlite occurs in vertical pipes up to 2,000 feet in diameter. These have been worked to depths of over 4,000 feet. Kimberlite is an ultrabasic rock, which is believed to have risen from great depths, possibly from the upper parts of the mantle. The diamonds are thought to have crystallized at great depths and to have been carried upwards by the ascending kimberlite. Diamond-bearing gravels are worked in Angola, Ghana, Brazil and India. In the Nama-qualand fields, the diamonds are found in uplifted beach deposits.

Graphite (C, No. 7. and Table p. 104) differs greatly in its physical properties from diamond, which is also pure carbon. Graphite is black to dark-grey in colour and is so soft that it marks paper. Its name is from the Greek word 'grapho', I write. It was formerly mistaken for lead, hence its alternative names 'plumbago' or 'black lead'. Crystals of graphite, which are rarely found, belong to the Hexagonal System, another difference from diamond, which is Cubic. Graphite is used in the so-called 'lead' pencils, as lubricant, for making crucibles, in the paint industry, especially for protective paints, and for a variety of electrical purposes (electrodes, dry batteries, etc.). For trade purposes, two classes of graphite, 'amorphous' and 'crystalline' are recognized, but these terms are not used in their strict mineralogical sense, but simply to denote differences in grain-size, 'amorphous' graphite being the finer grained.

Graphite normally occurs as flakes and lumps in a wide variety of igneous and metamorphic rocks, though in Ceylon there are veins up to 9 inches in width filled almost completely with graphite. Most deposits of graphite, as for example those of Korea, Ontario and New Mexico are the result of the regional or contact metamorphism of coal seams or of other carbonaceous sediments.

Sulphur (S. No. 268 and Table p. 105) occurs either in massive form as an en-crustation or as Orthorhombic crystals. Its typical sulphur-yellow colour is often tinged with red or green. The gases emitted by volcanoes are strongly sulphurous, so the rocks round the craters of many active and extinct volcanoes and hot springs are often encrusted with sulphur. In 1936 during an eruption in Japan, considerable flows of pure sulphur at temperatures of 120°C were discharged. The main commercial deposits of sulphur in Texas and Louisiana occur in the cap rock of salt-domes (Fig. 20), in the cavities of limestones or interbedded with gypsum. In Sicily sulphur is found disseminated through a cellular lime-stone and also in thin layers inter-bedded with gypsum. The origin of such deposits is puzzling, but an important part may have been played by certain bacteria, which can break down sulphides and sulphates to produce sulphur.

SULPHIDES

Pyrites (Iron pyrites, sulphur ore, FeS_2, Nos. 5, 8-12 and Table p. 105) is widely distributed, for it is the commonest sulphide in the earth's crust. If often occurs mixed with chalcopyrite or other sulphide minerals and is usually mined not as a source of iron, but for its content of sulphur or of other metals.

A colloquial term for pyrites is 'fools gold', but it can be distinguished by its more brassy hue, by the fact that it is only scratched by good quality steel (hardness 6-6.5) and its occurrence, if crystalline, as cubes, usually with striated faces (Fig. 15), or as the five-edged pyritohedron or pentagonal dodecahedron (Fig. 11, V).

The largest deposits of pyrites occur in south-west Spain. In the Rio Tinto mines, great lens-like bodies of almost pure pyrite are worked. They were probably formed by replacement of the country rock by hot mineralizing solutions. Other large pyrite deposits occur in Tasmania and in the Harz mountains of Germany, whilst through the world generally most deposits of sulphide minerals have a high pyrite contact. It also occurs in sediments, sometimes in concretions (No. 12), as in the Alum Shale of Sweden. Such sediments were deposited in seas where stagnant bottom waters became highly charged with hydrogen sulphi-de. The pyrites was formed partly by the action of the sulphur bacteria and partly by diagenetic changes. Small cubes of pyrites of metamorphic origin are common in many of the slates of North Wales.

Marcasite (No. 272) with the same chemical composition as pyrites, belongs to the Orthorhombic System and is of a paler yellow colour. Nodules of marcasite are found quite commonly in the lower beds of the Chalk. They usually have a brown oxidized crust and when broken show a radiating structure of yellow fibres, which soon tarnish on exposure to the air.

Pyrrhotite or Magnetic pyrites ($FenSn_{+1}$, Nos. 13 and 34, and Table p. 105) may contain up to 5% of Nickel and is indeed the most valuable ore of nickel. It is generally found massive, when crystalline it occurs as tabular Hexagonal prisms. On exposure it rapidly tarnishes from its fresh brownish or reddish colour. It is softer than pyrites (hardness 3.5-4.5) and is magnetic.

The largest deposits of pyrrhotite and other nickel sulphides occur at Sudbury in Ontario. The ore-bodies are either scattered along the base of a thick basin-shaped body of a gabbroic rock (norite) or in faulted zones. The ores used to be regarded as due to gravitational sinking of the heavy sulphides as the norite consolidated, but it is now believed that the ore bodies are later than the norite and that they were injected along faulted or brecciated zones with considerable replacement of the country rock.

Other occurrences of nickeliferous sulphides are in Finland, Sweden, in the Ural Mountains and in Manitoba.

Fig. 15. Striated cube of pyrites

Mispickel (Arsenical pyrites, FeAsS, No. 14, and Table p. 105) is the chief ore of arsenic. It is found associated with other sulphide ores. The arsenical content may cause great difficulties in the purification of the other metals. The arsenic is deposited in the flue dust of the smelters and extracted as a by-product. This is the case in the copper-lead mines of Butte, Montana and in the gold mines of western Australia and Brazil. At Boliden in northern Sweden the sulphide ore body contains up to 10% of arsenic. The amount produced exceeds the demand, so large quantities have to be stored.

Realgar (AsS, No. 269) and **Orpiment** (As$_2$S$_3$, No. 270, Table p. 105) occur in the oxidized parts of arsenic-bearing veins and also locally are deposited around hot springs (Steamboat Springs, Nevada) and around some volcanoes. Owing to their strong colours, (red and yellow respectively) they are used in the pigment industry.

Chalcopyrite (Copper pyrites, Cu$_2$S. Fe$_2$S$_3$, No. 16 and Table p. 106) is one of the chief ores of copper and occurs in many sulphide ore bodies. It is often round associated with iron pyrites, but can be distinguished by its less brassy colour, its tendency to tarnish and the fact that it can be scratched with a knife.

Bornite (Variegated copper ore or Peacock ore, a sulphide of copper and iron, No. 17 and Table p. 106) and *Chalcocite* (Copper Glance, No. 18 and Table p. 106) are often found in the same ore bodies, particularly in the zone of enrichment (p. 94).

Extremely large deposits of disseminated copper sulphide ores occur at Bingham in Utah, at Ely in Nevada, in Mexico and Chile and also in the famous Copper Belt of Katanga and

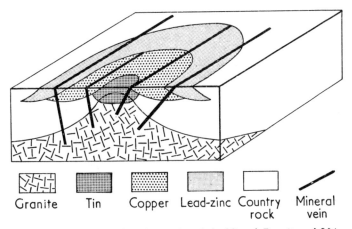

Granite   Tin   Copper   Lead-zinc   Country rock   Mineral vein

Fig. 16. *Block diagram to show the zonation of the Mineral Deposits and Veins of Cornwall.*

Zambia in Africa. The lodes of the largely worked out Cornish mines showed a well marked mineral zonation (Fig. 16). The deepest parts of the lodes yielded tin and some wolfram. This was overlain by a zone mainly of chalcopyrite, passing upwards into the zone of secondary enrichment, whilst lead, zinc and iron occurred near the surface. This mineral zoning, which was accompanied by a zoning of the gangue minerals, reflects the temperature at which the minerals formed. Tin at 550°C. or above, copper at 500°C, lead at 400°C. and the iron carbonates at temperatures as low as 150°C. A similar zonation is shown around the nearby granite masses. The copper and tin mines worked lodes in the granite, whilst the lead, zinc and iron mines were in the 'country rock' surrounding the granites.

Galena (Lead glance, PbS, Nos. 19 and 20 and Table p. 106) is the chief ore of lead. It is often found associated with zinc blende. Galena nearly always contains some silver. Most of the world's silver is obtained from argentiferous galena ores, though usually as a by-product to the lead. Galena usually occurs in lodes or as replacements of limestones. Famous lead mining fields occur at Broken Hill in New South Wales, in British Columbia (Sullivan Mine), in Burma, in the Leadville District of Colorado and in New Jersey. In Great Britain the lead-zinc lodes of the Mendips, Shropshire, Derbyshire, the Pennines and the Leadhills area in Southern Scotland have, in some instances, been worked from at least Roman times, but the output was never really large. Attractive specimens can still be collected from the spoil heaps of the disused mines.

Cinnabar (HgS, No. 271 and Table p. 106) is the chief ore of mercury or quicksilver. It occurs as disseminations, stock works or lodes in a variety of country rocks. The lodes often contain small globules of mercury. The deposits never extend to any great depth from the surface and are a product of hot aqueous solutions rising during periods of volcanic activity. Cinnabar is being deposited today around a number of hot springs in volcanic areas. The richest mine, the Almaden Mine in Southern Spain, has been worked for the past 2,000 years, whilst the mines in Tuscany, Italy and near Trieste date from the days of the Greeks and the Romans. In the New World, mercury deposits occur in California, Mexico and Peru.

Blende (Sphalerite, black jack, ZnS, Nos. 21-2 and Table p. 106) is the chief ore of zinc. As mentioned above, blende usually occurs associated with galena in lodes or disseminations of hydrothermal origin. Blende varies considerably in colour and therefore can be mistaken for other minerals. Indeed its alternative name Sphalerite is taken from the Greek word for 'treacherous'. Its resinous lustre and perfect rhombohedral cleavage are the features to look for.

Molybdenite (MoS_2, Nos. 23-4 and Table p. 107) is the chief ore of molybdenum. It is usually found associated with granitic rocks. The largest molybdenum deposit is at Climax, Colerado, where innumerable veinlets carrying molybdenite and gangue minerals, traverse altered granite. In some other mining fields in Mexico and Chile, molybdenum is present in small quantities in copper ores and is produced as a by-product. In Norway and Sweden, certain quartz veins carry small quantities of molybdenite.

Stibnite (Antimony glance, antimonite, Sb_2S_3, No. 25 and Table p. 107) is the chief ore of antimony. Its usual habit of lead-grey, somewhat irridescent, radiating crystals is very distinctive. It occurs either in lodes or as replacement deposits in limestone and shales and was formed from low temperature hydrothermal solutions. The chief mining areas are in China (where antimonite has long been used for darkening the eyelids), Bolivia and Mexico, but stibnite is occasionally to be found in the veins of Cornwall and other parts of Europe.

Cobaltite (CoAsS, No. 26 and Table p. 107), linnaeite (Co_2S_4) and smaltite ($CoAs_2$), are the common cobalt minerals. In the upper parts of ore bodies, they often oxidize to form erythrite, 'cobalt bloom', of a beautiful peach-red hue. Cobalt is obtained as a by-product from the mining of other sulphide ores. It was obtained from the silver ores of Cobalt in Ontario. The ores of the Copper belt of Katanga and Rhodesia carry cobalt, whilst in Morocco it occurs in gold ores and in Burma in nickel ores.

OXIDES

Quartz (SiO_2, Nos. 27-35 and Table p. 107) is the commonest of all minerals. It is an essential constituent of the more acid igneous rocks, it occurs in many sediments, especially the more clastic ones, in many metamorphic rocks and is a common gangue mineral. Many of the best specimens are found lining 'geodes' or 'vugs'. Quartz sands, sandstones and quartzite are composed almost entirely of water-worn grains of quartz. Their use as building sands, building stones, moulding sands, glass sands, etc., is determined partly by the size and grading of the quartz grains and partly by the nature and amount of the matrix to the quartz grains.

Pure quartz (No. 27) is colourless and water-clear. When first discovered in the Alps, such rock-crystal was thought to be frozen water and was named after the Greek word for 'ice'. Crystals and waterworn pebbles of rock crystal have often been labelled 'diamond', a mistake sometimes made deliberately. Quartz can, however, be easily distinguished from diamond by its crystallography, for quartz is Trigonal and not

Cubic, by its lower degree of lustre and by its inferior hardness. Rock-crystal is used in the jewellery trade and for making the best quality glass. It also possesses piezo-electric properties, that is flawless crystals of clear quartz expand and contract very slightly when subjected to an electric field. They are used for quartz-resonators and oscillators in radio and telecommunications. Much of the best quality quartz for these purposes comes from Brazil. There are many coloured varieties of quartz. Amethyst (No. 32), purple or violet in hue, is used in jewellery. Again many of the best crystals come from Brazil. Rose quartz (No. 33) is apt to lose its colour on exposure, but if moistened this will reappear. Cairngorm, found in the Cairngorm Mountains of Scotland and used in Highland jewellery, is a rather pale variety of Smoky quartz (Nos. 30-1), whilst the name morion is given to almost black stones. Milky quartz (No. 29) is due to innumerable very small air bubbles in the quartz. Citrin (Nos. 34-5) is the yellow variety of quartz. Aventurin quartz contains spangles of mica or specular iron ore, whilst ferruginous quartz is coloured by the iron oxides present. Tiger's eye (No. 38) is silicified crocidolite (p. 137), an asbestiform mineral.

Opal (amorphous silica, No. 41 and Table p. 107) is hydrated silica with formula $SiO_2 \cdot nH_2O$, with usually less than 10% of water. It can be regarded as a form of solidified jelly.

The gem variety *Precious Opal,* shows a magnificent play of colour, due to the presence of extremely thin films, which cause refraction of the light falling on them. The colour of the opal ranges from black (rare) to white. Reddish or orange tinged varieties are called fire-opal. The richest opal fields are on the boundary of New South Wales and Queensland in Australia, where the opal is found in sandstones. The opals used by the Romans, probably came from Czechoslovakia, where opal occurs in the crevices of andesites near hot springs. Many beautiful fire-opals have been obtained from Mexico.

Fossils are sometimes replaced by opal, especially wood opal. Deposits of hydrous silica, or siliceous sinter are being deposited round many hot springs and geysers in such active volcanic regions as New Zealand, Iceland and parts of the United States.

Chalcedonic Silica (Nos. 36-40 and 43-4, Table p. 107) is a mixture of quartz and opal, that is of crystalline and amorphous hydrated silica. It occurs as three main varieties:- Chalcedony, Flint and Jasper.

Chalcedony is found usually either infilling steam cavities in lavas or more rarely in lodes. According to its colour, a large number of varieties have been recognized, such as carnelian, No. 36 (reddish or reddish-brown), plasma, (bright green speckled with white), bloodstone, No. 40, (speckled with red), agate, No. 37, (showing marked colour banding), moss agate (with small dendritic growths of iron oxide) and onyx (with flat sided colour bands. Some of these varieties, especially agate, have been used for ornamental purposes.

Flint (No. 43-44) is compact cryptocrystalline silica, which occurs in layers and irregular shaped nodules in the upper part of the Chalk. Water-worn pebbles of flint are widespread in the river gravels and beaches of southern England. The river gravels are worked extensively as aggregates for concrete. In the limestones and sandstones of other parts of England, notably in Derbyshire and Dorset, occur beds and masses of *Chert.* Flint is normally grey to black in colour, chert often rather browner. Flint, when struck breaks with a curved (conchoidal) fracture, chert with a rather flatter fracture, but it is often extremely difficult, if not impossible, to distinguish flint from chert. Flint also occurs in the Chalk of other parts of Europe, North France, Belgium, Denmark, etc., whilst chert is often found in limestone of many parts of the world. Man's first tools were made from such hard, yet brittle rocks, as flint and chert. Flakes can be easily struck off a flint nodule, but to be sure that the flake will be of the shape required, requires much skill and practice.

Jasper (No. 39) is an impure form of opaque cryptocrystalline silica, usually red or brown in colour. Sometimes as in ribbon jasper it shows beautiful colour banding.

Rutile (TiO_2, No. 45 and Table p. 107) is one of the chief ores of titanium.

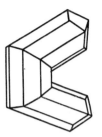

Fig. 17. *Geniculate (knee shaped) twin of rutile.*

Rutile occurs as an accessory mineral in many igneous and metamorphic rocks. Workable deposits are rare. In Virginia, U.S.A., rutile occurs with ilmenite ($FeTiO_3$) in a feldspathic pegmatite vein, whilst at Kragero in Southern Norway rutile is disseminated in an aplite. Rutile is also a detrital mineral in arenaceous sedimentary rocks, usually in very small amounts, but in the modern beach sands on the east coast of Australia it is concentrated in workable amounts.

Cassiterite (Tinstone SnO_2, No. 46 and Table p. 108) is the chief ore of tin. It occurs in lodes associated with acid igneous rocks in Cornwall, Malaya, Bolivia, Nigeria and China. As already described on p. 122, tin in Cornwall occurs only in the deepest parts of the lodes. Cassiterite is a heavy mineral (S.G. 7) and the bulk of the world's output of tin, comes not from the mother lodes, but from alluvial and eluvial deposits, notably those of Malaya and Indonesia. Stream tin, water-worn cassiterite, can still be found in the river gravels and beach sands of Cornwall.

Cuprite (Cu_2O, Table p. 108) occurs in the zone of oxidized enrichment (Fig. 9) of many copper mining fields. Its red colour is very distinctive. The name chalcotrichite (after the Greek words for copper and thread) is given to a variety consisting of delicate interlacing fibrous crystals of a beautiful cochineal-red hue.

Magnetite (Magnetic or Black iron ore, Fe_3O_4, Nos. 52-4, Table p. 108) is an accessory mineral in many igneous rocks and where it occurs in sufficient concentration it is a very valuable ore of iron. Very rich deposits are found at Kiruna in northern Sweden. The steeply dipping main ore body up to 500 ft. in width, has a 'hanging wall' (upward facing contact) of syenite porphyry and a 'foot wall' (downward facing contact) of quartz porphyry. It is generally regarded as a magnetic differentiate at depth, which was injected as a molten sheet along the contact of the two porphyries. Owing to its high phosphorous content, this rich ore body could not be worked until suitable methods of removing the phosporus had been developed. Other magnetite ore bodies due to contact-metamorphism, occur in the United States and in the Urals. Magnetite is locally concentrated in placer deposits and beach sands.

Lodestone is the name given to magnetite with natural polarity. The earliest form of marine compass consisted of a splinter of lodestone in a bowl of water.

Hematite (Specular iron ore, kidney ore, Fe_2O_3, Nos. 50 and 273, Table p. 108) is another very important ore of iron. Its name, after the Greek word for blood, is due to its very distinctive colour. In the Furness district of Cumberland and in the Forest of Dean, bodies of hematite, often very irregular in form, occur in limestones. By

125

Fig. 18. *Kidney iron ore (hematite)*

some they are regarded as formed by hot mineralizing solutions rising from below and replacing the limestone, by others the iron is believed to have originated from ferruginous sandstones, which overlie or formerly overlaid the limestones. The hematite ores of Bilbao in Spain are due to the oxidation of original carbonate ores. The world's most productive iron mines are around Lake Superior. They are bedded deposits consisting of interbedded layers of chert and of haematitic ore (p. 163). Other important sedimentary iron ore beds occur at Clinton in Alabama, where the hematitic ore often shows an oolitic structure, (p. 166) and in the Ukraine, where the hematite is in the form of black octahedra, martite, and is probably pseudomorphic after magnetite.

As well as martite, there are other varieties of hematite. Specular iron ore (looking-glass ore), black rhombohedral crystals with splendent metallic lustre, used in the jewellery trade, micaceous hematite occurs in thin scales, kidney ore (Fig. 18) is a reniform variety and reddle, an earthy variety.

Corundum (Al_2O_3, No. 47 and Table p. 108) is an important abrasive, for it is the next hardest mineral after diamond. Ordinary corundum is usually found as barrel-shaped Trigonal crystals (Fig. 19) which may be much

water-worn in alluvial deposits. Corundum occurs in several different geological settings. As magmatic segregations in nepheline syenites (Ontario), along the margins of pegmatite dykes (the Transvaal) or at the contact between ultrabasic intrusions and gneissose country rock (Southern Appalachians). *Emery* is a greyish-black variety of corundum, admixed with magnetite and hematite. Whilst still a good abrasive, it is not as hard as pure corundum. It occurs as segregations in igneous rocks or granular limestones at Noxos in Greece, in Turkey and in the Urals.

Common corundum is dull and rather colourless, but there are other varieties which form superb gemstones. *Sapphire, ruby* and less attractive stones in shades of yellow, green, violet, etc., to which the names oriental-topaz, oriental-emerald and oriental-amethyst have been given by jewellers. The red colouration of rubies (No. 48) is due to the presence of minute quantities of chromium, the blue colour of the sapphire (No. 49) to the presence of titanium. The world's supply of fine rubies has come almost entirely from the Mogok Mines in Upper Burma, where rubies occur with sapphires, spinel, tourmaline, beryl, garnet, etc. in contact-altered limestones. The alluvial deposits of the same areas has yielded many fine gems. Fine sapphires are found in the famous gem gravels of Ceylon, and also in Siam and Kashmir. Good quality sapphires occur in a much weathered dyke in Montana and also in Queensland and New South Wales.

Spinel (MgO, Al_2O_3, No. 277, Table p. 108) occurs as an accessory mineral in basic and ultrabasic igneous rocks and in certain contact-altered shales and limestones. Spinel of gem quality is found in many of the areas which yield sapphires and rubies. Ruby-spinel, the clear red variety, has often been confused with the true ruby. Indeed the so-called Black Prince's Ruby, worn by Henry V at the battle of Agincourt and now very conspicuously placed in the British Imperial State Crown, is really a 'balas-ruby' a rose-red spinel. Other varieties of spinel are pleonaste, (dark-green and containing much iron), black hercynite and brown picotite, containing iron and chromium.

Braunite (Mn_2O_3, No. 57, and Table p. 109) and *Manganite* ($Mn_2O_3H_2O$, No. 62, and Table p. 109) occur occasionally in lodes but more often as residual deposits, formed by the weathering of other manganese minerals.

Pyrolusite (MnO_2, No. 274-5 and Table p. 109) is, however, a much more important ore of manganese, occuring mainly as residual masses in the U.S.S.R., in the central Provinces of India, near Kimberley in South Africa, Brazil and elsewhere. It usually occurs in masses, sometimes showing a fibrous habit. In England joint surfaces in the Chalk often show dendritic markings due to thin films of dark pyrolusite.

Limonite (Brown hematite, hydrated iron oxide, $2Fe_2O_3.3H_2O$, No. 60-1 and Table p. 109) is formed from the other iron minerals. The bedded iron ores of the southern Midlands of England and of the Lorraine fields in France, often have a yellow or brown limonitic crust. When this is broken with a hammer, the unweathered green carbonate-silicate ore is exposed. It forms the 'gossan' or iron cap (Fig. 9) over veins and bodies of sulphidic ores containing iron pyrites. In more tropical climates, as for example, Cuba, there are rich residual deposits, mainly of limonite, in the lateritic soils. Limonite is being precipitated, partly by bacterial action, on the floors of lakes in Sweden to form *bog iron ore*. In one case a layer of 7 inches in thickness accumulated in 26 years. Penny ore (No. 59) is the name given to one particular form assumed. Other variants are called bean ore, pearl ore, etc.

Ilmenite (FeO. TiO_2, No. 58 and Table p. 109) is an accessory mineral of the more basic igneous rocks and may form large lens shaped segregations, as in the Egersund district of Southern Norway and at Tabery in Sweden. These titanifeours iron ores used to be worked for iron, but more recently the emphasis has changed to the production of the paint titanium white. Similar deposits occur in the Adirondack Mountains of the eastern United States and in Quebec. The beach sands at Travancore in South-west India are worked for their high content of ilmenite (up to 70%). These beach sands also contain unusual amounts of zirconium and cerium minerals.

Fig. 19.   *Barrel shaped crystals of corundum.*

Rock Salt (Halite, common salt, NaCl, No. 63 and Table p. 109) is formed from the evaporation of bodies of seawater. Seawater contains about 3.5 % of dissolved salts, mainly sodium chloride (2.7 %) whilst the remaining 0.8 % is made up of compounds of magnesium, calcium and potassium. If a body of seawater is evaporated to dryness, its contained salts are precipitated in order of their solubility, the most soluble salts last. A stratified deposit will be formed with the layers arranged in the following order from least soluble at the base to most soluble at the top:-

4. Potassium-rich layer (polyhalite, sylvine, carnallite, etc.)
3. Rock salt layer
2. Anhydrite layer
1. Dolomitic limestone layer.

If the conditions permit steady replenishment of saline water to replace the salt that is precipitated, then beds

Fig. 20 *Section through a salt dome. A younger bed rests unconformably on the strata uparched by the rising salt. Note the plastic folding of the salt within the dome and also the accumulation of oil against the sides of the dome in the beds that are cross-patched.*

of rock salt many feet in thickness may be laid down. At Winsford in Cheshire a bed of rock salt, 60 feet in thickness, is mined. This thickness is often exceeded in other parts of the world.

Salt deposits of this kind form part of a succession of sedimentary rocks and may in the course of time be overlain by thousands of feet of other rocks. When heavily loaded by the weight of these overlying rocks, the salt bed becomes plastic and flows slowly into regions of lower pressure. In this way *salt domes* are produced (Fig. 20), plug-shaped bodies of salt, up to several miles in diameter and thousands of feet in thickness. The salt domes are formed not only of halite, but also contain layers of the potassium and magnesium-rich salts. The beds surrounding the salt domes are uparched and may form traps in which mineral oils accumulate. Such salt domes occur in Persia, Texas and north Germany. The north Germany salt domes are extensively exploited, both for the salts, particularly the potash salts, which they contain in vast quantities, and also for the mineral oils obtained from the surrounding rim of uparched sediments. In the Texan oil fields, the cap-rock overlying the salt domes yields appreciable quantities of sulphur.

Bedded salt deposits, not salt domes, occur at depths of up to a few thousand feet, beneath Cheshire, Staffordshire, east Yorkshire, the Stassfurt area in north Germany, Salzburg in Austria, in the area south of the Great Lakes in the U.S.A., and elsewhere. The salt is extracted either by normal mining methods or by 'brining', boring to the salt bed, forcing down steam to dissolve the salt and pumping the resultant brine to the surface.

Fig. 21.   *Hollow faced cube of halite.*

As well as beds of massive halite, one can sometimes find beds containing 'salt pseudomorphs', that is cubes of clay after salt, with hollow faces (Fig. 21). Evaporation of a body of sea water caused the precipitation of cubes of halite. An inflow of water, carrying mud or silt, dissolved away the rock salt and the space left was infilled with mud or silt to form a pseudomorph. Exposure to the sun would then harden the infilling sufficiently for it to retain its shape, when it was buried beneath later beds. Such salt pseudomorphs can be found at Pinhay Bay near Lyme Regis, Dorset, at Watchet on the Somerset coast and in other localities. They occur in beds which do not contain thick layers of salt, but must have been laid down under semi-arid conditions with many temporary pools, conditions similar to those of the 'vleys' of the semidesert regions of South Africa. Thick beds of salt and gypsum are now being deposited in the Dead Sea, in the Great Salt Lake of Utah and in Lake Eyre in South Australia.

Fluorite (Fluor Spar, CaF_2, Nos. 64-5 and Table p. 109) is one of the chief gangue minerals in lead-zinc lodes of Derbyshire and in the Alston area of the Pennines. It is a less common gangue mineral in the copper-tin lodes of Cornwall, but may occur in appreciable quantities in some of the pneu-matolitically altered granites of south-west England, e.g. china stone (p. 154). Fluorite can occur in a wide range of colours. The deep blue or purple variety known as Blue John or Derbyshire spar, used to be mined at Castleton in Derbyshire and carved into vases, etc. The fluorite in china stone is usually a less intense shade of purplish blue than Blue John. Fluorite also occurs in a wide variety of other colours, white, green, yellow, etc. Whatever the colour, it can be recognized by its form, cubic crystals with perfect octahedral cleavage (Fig. 11, II) and by its hardness (4). Hydrothermal veins and replacement deposits of fluorite occur in many other regions of sulphide ores, for example in the Harz Mountains of Germany. In the Illinois-Kentucky area of the United States, lodes composed almost entirely of fluorite are up to 35 feet in width.

Many specimens of fluorite are fluorescent, that is they glow with a lovely violet light, when exposed to ultra-violet light. This property is due to the presence of minute amounts of manganese and the rare-earth minerals.

Cryolite (AlF_3. $3NaF$, No. 66, Table p. 109) is a rare mineral, only found in quantities worth mining at Ivigtut on the west coast of Greenland, where it occurs in a pegmatite dyke cutting granite. It is associated with some sulphide ores and a number of rare minerals. The name cryolite (from the Greek word for 'frost stone') was given from its resemblance to ice. A piece of cryolite disappears if it is placed in water, for it has practically the same refractive index as water. It will be recalled that H.G. Wells' unfortunate 'Invisible Man' disappeared because the refractive index of his skin had been changed to that of air. At one time cryolite was the only known source of aluminium, but it has now been replaced by bauxite (p. 169), though cryolite is still required for certain purposes in the aluminium industry.

Apatite (phosphate of calcium, with fluorine and chlorine, Nos. 82-3 and Table p. 110) is an accessory mineral of igneous rocks. It is present in small amounts in many metamorphic rocks, especially in metamorphosed limestones: Pegmatites, containing apatite in workable quantities, occur in Ontario, Quebec, and in Norway, whilst in the alkali-syenites of the Kola Peninsula, Northern Russia, there are great lenses of apatite-neheline rock which are worked as a source of phosphate. Apatite crystals are often attractive in appearance and have been used in the past as gemstones, but their hardness (5) is too low for these 'gems' to stand much wear, hence the mineral's name after the Greek word for 'deceit'.

CARBONATES

Calcite (Calc Spar, $CaCO_3$, Nos. 67-72 and Table p. 110) is a widely distributed mineral, occurring in igneous, metamorphic and sedimentary rocks, and also as a common gangue mineral in vein deposits. Well shaped crystals are quite commonly found. They are Trigonal with the hexagonal prism well developed. In nail-head spar (Fig. 12 III) the prisms are terminated by flat rhombohedra, in dog-tooth spar (Fig. 12, II) by scalenohedron faces. At first sight, colourless calcite may be confused with quartz, but it is distinguished by its hardness (3) and by its effervescence with dilute hydrochloric acid. Calcite has a perfect rhombohedral cleavage. *Iceland Spar* is a very pure transparent variety of calcite, which shows the property of double refraction (No. 67). Because of this, iceland spar used to be used for the nicol prisms of petrological microscopes, but it has now been replaced by 'polaroid'. Twin crystals of calcite (No. 68) are not uncommon. Calcite is commonly colourless, but often shows other hues, especially when it has been stained by iron (Nos. 71-2). Fibrous calcite with a satin-like lustre is known as 'beef', an old quarryman's term. 'Beef' occurs in sedimentary rocks, particularly limestones, as narrow veins usually along the bedding, the fibres lying at right angles to the walls of the veins.

Calcite is an important constituent of many sedimentary and metamorphic rocks, (stalactites, tufa, chalk, limestone, marble, etc.) which are dealt with in the next section.

Aragonite, (No. 73 and Table p. 110) has the same chemical composition as calcite. It differs from calcite in being Orthorhombic, usually crystallizing as sharp pointed prismatic crystals. Twinning is common, the twins having a pseudo-hexagonal shape, but distinguishable owing to be presence of reentrant angles. Aragonite can also be distinguished from calcite by its greater hardness (3.5-4), it is slightly heavier, whilst it is poorly cleaved and the cleavage is parallel to pinacoidal and not rhombohedral faces.

Aragonite is less stable than calcite into which it changes, owing to the effect of heat, pressure or time. Many organisms secrete their skeletons of aragonite, but these are only found unchanged in rocks of relatively recent geological age. In older rocks these shells have either altered to calcite or have become unrecognizable.

Dolomite (Pearl Spar, $MgCO_3.CaCO_3$, No. 278 and Table p. 110), occurs as rhombohedral crystals (Fig. 12, I) but the faces are usually curved. It also possesses a perfect rhombohedral cleavage. Sometimes white, dolomite is more often honey coloured. Unlike calcite with which it may be confused, dolomite does not effervesce with cold dilute hydrochloric acid, whilst the crystal faces have a pearly, rather than a vitreous lustre. The name pearl spar is given to the variety that is often found as a gangue mineral, commonly associated with blende or galena.

Dolomite is also an important constituent of many sedimentary rocks, occurring either as the basal member of an evaporite sequence (p. 128) or in dolomitic limestones (p. 167).

Magnesite ($MgCO_3$, Table p. 111) is the chief ore of magnesium. It is rarely found crystalline but it can form rhombohedral crystals strongly resembling those of dolomite. More commonly it is either massive and fibrous, or granular and chalk-like. It occurs either as ramifying veins or a stockwork in serpentinite (p. 93), as in the Isle of Euboea, Greece, or as in replacement deposits in limestone or dolomitic limestones. Some of these replacement deposits, as in Manchuria, the Ural Mountains and near Chewelah, Washington, U.S.A., are extremely large. The vein deposits are regarded as having been formed by carbonated waters, which have reacted with the magnesia-rich serpentinite, replacement deposits were probably due to magnesia-bearing solutions enamating from an igneous magma.

Siderite (Chalybite, Spathose iron, $FeCO_3$, No. 74 and Table p. 111), occurs both as a vein-mineral and as an important constituent of certain sedimentary iron ores (p. 162). When crystalline it occurs as curved rhombohedra, yellowish or brown in colour, distinctly heavier than dolomite (S.G. 3.8 as against 2.8) It is also found massive and granular.

Malachite ($CuCO_3. Cu(OH)_2$, No. 75 and Table p. 111) is an important ore of copper, being one of the typical minerals of the zone of secondary enrichment (Fig. 9). Owing to its bright green colour it is easily recognizable. It normally occurs as massive, encrusting or stalactitic masses with a smooth mammilated surface. Malachite is something used for ornamental purposes, for when cut and polished it shows a most attractive colour banding.

Azurite (Chessylite, $2CuCO_3.Cu(OH)_2$, No. 76 and Table p. 111) is another important copper ore, often found associated with malachite in the zone of secondary enrichment (Fig. 9). If crystalline, it occurs as Monoclinic prisms, but commonly it is massive. The beautiful azure-blue colour of the mineral is very distinctive.

Witherite ($BaCO_3$, Table p. 111) is found as a gangue mineral, associated with galena and barytes, in the mineral veins of the northern Pennines. When crystalline it occurs as white or greyish, Orthorhombic crystals, which have been twinned to produce six-sided prisms, rather like quartz, but with re-entrant angles, whilst their hardness is 3.5 instead of 7.

SULPHATES

Barytes (Barites, Heavy Spar, $BaSO_4$, No. 77 and Table p. 111) is a white, yellowish or sometimes blue-tinged mineral, which can often be recognized by its weight (S.G. 4.5). It occurs as flat Orthorhombic crystals, (Fig. 11, VIII) which in the variety known as cockscomb barytes form tabular masses. The crystals show a perfect basal and prismatic cleavage and a resinous to pearly lustre. Barytes also occurs in massive, granular, columnar and sometimes stalactitic forms. It is found in many parts of the world, as a gangue mineral in lead-zinc veins. In Georgia and Virginia, U.S.A. there are important residual deposits, consisting of barite nodules set in a clay and derived from the weathering of primary deposits.

Barytes has a considerable number of economic uses, ranging from an important constituent of white paint, an inert filler of paper, linoleum, plastics and face powder, etc., and in tanning.

Gypsum ($CaSO_4.2H_2O$, Nos 78-80 and Table p. 111) in its crystallized variety *Selenite,* occurs as Monoclinic crystals, often as swallow-tail twins (Fig. 22). The crystals frequently occur as stellate, interpenetrating groups. Apart from their crystallography, such crystals have a number of other distinctive features; first they are easily scratched with the fingernail (H 2), secondly gypsum has a perfect pinacoidal cleavage and these cleavage faces show a beautiful pearly lustre (No. 78). Gypsum also occurs massive, and if fine-grained enough forms *Alabaster,* which has long been used for ornamental purposes. *Satin spar* with its silky lustre is the fibrous form of gypsum.

Bedded deposits of gypsum usually form part of evaporite sequences and are worked on a large scale in England, Germany, and the United States and elsewhere. Calcined gypsum is widely used in the building trade for plasters, cements, wall boards, etc. Other uses are as a filler for paper, paint, etc. 'Plaster of Paris' is formed by heating gypsum to about 350°F. to drive off one and a half molecules of its water of crystallization. Good crystals of selenite are to be found in many clays, particularly those of Mesozoic and Tertiary age. They have been formed from the decomposition of pyrites, present in the clays, producing sulphuric acid, which reacts with calcareous fossils to form calcium sulphate.

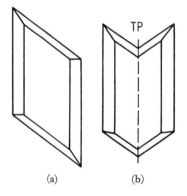

Fig. 22. *Gypsum crystals; (a) Simple monoclinic crystal of selenite, (b) Swallow tail twin of selenite with twin plane shown by broken line.*

Anhydrite ($CaSO_4$, Table p. 112) is another member of the evaporite cycle (p. 128). In evaporite deposits, and particularly in salt domes, anhydrite and gypsum are often intimately inter-mixed. It is probable that the anhydrite has been irregularly converted to gypsum.

Anhydrite is harder than gypsum (H 3.5 as against 2), somewhat heavier (S.G. 2.9 compared with 2.3) and crystallizes as tabular Orthorhombic crystals, which break up into rectangular fragments, owing to the presence of three perfect cleavages, as compared with the one perfect cleavage of gypsum. Anhydrite also occurs fibrous, granular and compact. When in the massive form, anhydrite tends to be very pale blue in colour.

WOLFRAMITES

Scheelite ($CaWO_4$, No. 81 and Table p. 112) is an important ore of tungsten. It occurs, associated with tinstone, topaz, etc. and other pneumatolytic minerals in veins in Cornwall, Cumberland, the Harz Mountains, Arizona, etc. Scheelite, when crystallized forms white, yellowish or brownish lustrous Tetragonal crystals. More commonly it is found in reniform masses with a columnar structure but is sometimes massive and granular.

SILICATES

The *Feldspars* are a most important group of rock forming minerals. They are an important constituent of most igneous, many metamorphic and certain sedimentary rocks. They can be divided into the Alkali Feldspars (Orthoclase and Microcline) and the Plagioclase Series, the Soda-Lime Feldspars.

Orthoclase ($KAlSi_3O_8$, Nos. 84, 85 and 87 and Table p. 112) occurs as white or flesh coloured Monoclinic crystals (Fig. 11, IX) which often show simple twinning (Fig. 12, VIII). Its hardness is 6, so it is just scratched by a knife, unlike quartz. Orthoclase also has a duller lustre than quartz and indeed may appear almost opaque. These distinctions are important, for particularly when examining igneous rocks, it is essential not to confuse small crystals of quartz with feldspar. *Moonstone* is an opalescent to pearly variety of Orthoclase.

Orthoclase is an essential constituent of the more acid igneous rocks. In some pegmatites, crystals up to four or five feet in length are not uncommon. Such pegmatites are worked commercially in the United States, Canada, Sweden, Norway, etc., for feldspar is used in the ceramic industry for making opalescent glass, glaze, etc. Orthoclase also occurs in many metamorphic rocks and as a detrital mineral in arkosic sandstones (p. 157).

Microcline (No. 86 and Table p. 112) has the same chemical composition as orthoclase, but is Triclinic, but only just! In orthoclase the angle between the basal and the side pinacoid is 90°, in microcline 89° 50′, a distinction that very obviously cannot be detected in the hand specimen. Crystals of pinkish microcline and of pinkish orthoclase are therefore indistinguishable, except that there are two sets of fine striations crossing at right angles on the basal pinacoid of microcline. This is the result of repeated twinning. Much more easy to recognize is the bright green variety of microcline, known as *Amazonstone,* which is found in certain pegmatites in Colorado, Kashmir, the Urals, but curiously enough, not in the region of the Amazon River in South America. Like orthoclase, microcline occurs in acid igneous rocks, but is distinctly less common.

The Plagioclase or Soda-Lime Feld-

spars (Nos 88-9 and Table p. 112) form a series with continuous gradation in composition and physical properties between the two end members, albite, $NaAlSi_3O_8$, and anorthite, $CaAl_2Si_2O_8$. This isomorphous mixture of the albite and anorthite molecules is arbitrarly divided into a number of members, whose composition in terms of the Ab (albite) and An (anorthite) molecule is shown below:-

				S.G.
Albite	$Ab_{100}An_0$	$\rightarrow$	$Ab_{90}An_{10}$	2.605
Oligoclase	$Ab_{90}An_{10}$	$\rightarrow$	$Ab_{70}An_{30}$	2.649
Andesine	$Ab_{70}An_{30}$	$\rightarrow$	$Ab_{50}An_{50}$	2.660
Labradorite	$Ab_{50}An_{50}$	$\rightarrow$	$Ab_{30}An_{70}$	2.710
Bytownite	$Ab_{30}An_{70}$	$\rightarrow$	$Ab_{10}An_{90}$	2.733
Anorthite	$Ab_{10}An_{90}$	$\rightarrow$	Ab_0An_{100}	2.765

The variation in specific gravity is given as an example of the regularity of gradation in properties along the plagioclase series.

The plagioclases are Triclinic, the crystals resembling orthoclase (Fig. 11, IX) in being made up of pinacoids and hemiprisms, but the angle between the basal and the side pinacoids is 86° to 87° instead of 90°. The distinctive feature of the plagioclases is their repeated twinning (Fig. 12 IX). The traces of the parallel twin planes can usually be seen with a hand lens, and sometimes even with the unaided eye, on plagioclase crystals in rocks. The distinction between an orthoclase and plagioclase feldspar and also the precise position of a plagioclase in the plagioclase series can be carried out much more precisely in thin-section under the petrological microscope. In colour the plagioclases are usually white, except that *labradorite,* often but not invariably, shows schillerization – a rich play of colours, mainly in blues and greys (No. 89), due to the presence of myriads of minute rod-like inclusions,

apparently of iron-ore, that lie parallel to one another. *Moonstone,* a semi-precious gemstone, shows the same sheen to an even greater degree. In this case it is due to the intimate arrangement of thin layers of albite and orthoclase. The plagioclases are an essential constituent of the more basic igneous rocks, contrasting with orthoclase, which occurs in the more acid. Anorthosite (p. 156) is composed almost entirely of labradorite. Plagioclases also play an important part in many metamorphic rocks, particularly the gneisses (p. 172), schists (p. 171) and hornfelses (p. 172). Plagioclase, in rocks, is liable to alteration and may be partly or completely replaced by secondary white mica, scapolites, zeolites, epidotes or even calcite. These changes are known as *saussuritisation.*

The **Feldspathoids** are minerals, which are related to the feldspars. They are composed of the same elements, though in different proportions and are notably poorer in silica. They are found in certain syenites, basalts, etc.

are often referred to as the undersaturated rocks for they are low in silica and rich in alkalies.

Leucite ($KAlSi_2O_6$, No. 90 and Table p. 112) occurs as whitish or greyish crystals. They are Cubic and if well formed icositetrahedra (Fig. 11, VI) are eightsided in cross-section. With a good hand-lens symmetrically arranged inclusions may be visible. Leucite is only found in lavas. Certain of the flows on Vesuvius yield perfectly shaped crystals. Other localities are the Leucite Hills, Wyoming and the Kimberley District in Australia.

Nepheline (eleolite, $NaAlSiO_4$, No. 91 and Table p. 113) occurs as white, yellowish or brownish crystals, which are Hexagonal and, if well-formed are six-sided in cross-section. Ordinary nepheline occurs in undersaturated lavas as small glassy crystals, whilst the darker coloured variety found in certain syenites is known as *eleolite*.

Lapis-Lazuli (No. 276) is a rock composed mainly of a blue mineral hauyne, pyrites and calcite. The mines in the Badakshan district of Afghanistan which have been worked for 6,000 years, supplied the ancients both with a much prized gemstone and also with the powder for the pigment ultramarine. Other localities are in Siberia, Chile, Upper Burma and in California. The blue mineral, *hauyne,* or hauynite, a complex soda-lime aluminium silicate, also occurs in certain undersaturated lavas, as for example at Vesuvius and in the Eifel district of West Germany. There it occurs as small Cubic crystals, octahedra and rhombic dodecahedra, bright blue or greenish-blue in colour.

The **Pyroxenes** (Nos. 107-110 and Table p. 114) are an important group of rock-forming minerals with the general formula $R SiO_3$, where R may be calcium, magnesium or iron, but in detail their composition may be much more complex. They form a number of Series. For our purpose they can be divided into:-

1. The Orthorhombic Pyroxenes e.g. Enstatite
2. The Monoclinic Pyroxenes
   (a) The Diopside Series
   (b) The Augite Series
   (c) The Alkali-Pyroxene Series

The members of these series can be distinguished much more easily and accurately under the petrological microscope than in hand specimen. The various pyroxenes are important constituents of the more basic igneous rocks and of many metamorphic rocks. Perfectly shaped crystals of augite can sometimes be found in crystal-tuffs, whilst the more sodic pyroxenes are a distinctive feature of 'skarns' (p. 173) and other contact altered rocks.

Enstatite (No. 107) has the general composition $(Mg, Fe)O. SiO_2$. If not massive and lamellar, it occurs as stout prismatic Orthorhombic crystals grey, green, brown or yellow in hue. The variety bronzite has a pearly metallic lustre, whilst hypersthene, much richer in FeO. SiO_2 is brownish or almost black in colour, but often shows schillerization (p. 134).

Diopside (No. 108, $CaO. MgO. 2SiO_2$) when crystalline, forms prismatic Monoclinic crystals, white or dark greenish in colour. Chrome-diopside (No. 110) with a few per cent of Cr_2O_3 is a bright green variety.

Augite (No. 109), a complex silicate of calcium, magnesium, iron and aluminium, forms dull black Monoclinic crystals, which can be distin-

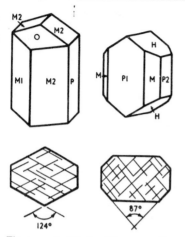

Fig. 23. *Amphibole and Pyroxene; Above Side views of simple crystals of hornblende (left) and of augite (right) H. - Faces of hemibypyramid) M - faces of prisms O. - Faces of hemiorthodome P. - Faces of pinacoids. Below Basal sections showing the two cleavages.*

The **Alkali-Pyroxenes** (acmite and aegirine) are in general similar in hand-specimen to the augites, except for the variety *Jadeite* $NaAl(SiO_3)_2$, which in the East is a highly prized ornamental stone. It is extremely tough, owing to the complex interweaving of individual crystals. It occurs in a variety of colours, but shades of green are the most highly prized. With hardness 7, they require great patience to work, but when polished have an attractive dimpled appearance. The finest jade comes from the Myitkyina district of Upper Burma, where the jade occurs as boulders in river terraces, or in the stream beds. The mines have been worked for a very long period.

The **Amphiboles** are another group of important rock-forming minerals. Like the pyroxenes, they are complex silicates of iron, magnesium and calcium, sometimes with sodium and/or aluminium. They resemble the pyroxenes, in many features. The chief distinctions in hand specimen, are tabulated below, but precise identification is made with much greater certainty under the petrological microscope. The chief members of the Amphibole family are briefly described below :-

guished from those of hornblende by their crystallography (Fig. 23). Twinned crystals with their characteristic re-entrant angle are not uncommon.

	AMPHIBOLES	PYROXENES
Angle between prism faces	120°	90°
Angle between cleavage	124°	87°
Horizontal crossection	six-sided	eight-sided
Crystals terminated by	three faces	two faces
Bladed forms	common	uncommon

Anthophyllite (No. 114, (Mg, Fe) $O.SiO_2$) characterizes certain metamorphic rocks which have been derived from basic or ultrabasic igneous rocks. It is Orthorhombic, but commonly occurs as prismatic needles and

radiating fibres, brownish in hue and of a vitreous lustre.

Tremolite-Actinolite Series vary in composition from the calcium-magnesium-silicate (tremolite) end to the calcium-magnesium-iron silicate (actinolite) end. They are commonly of fibrous habit or as long slender Monoclinic crystals. Tremolite (No. 115) is typically white or grey in colour, whilst the more iron-rich members towards the actinolite end are greener (No. 116). This series occurs in metamorphic rocks derived from the contact alteration of impure calcareous rocks or from the regional metamorphism of basic and ultrabasic igneous rocks.

There are a number of varieties of considerable economic importance. *Nephrite* or Greenstone is the commoner of the two jades. (For Jadeite see p. 136). It is composed of fibrous crystals, mottled or foliated together. Prehistoric implements made of nephrite have been found in Mexico and in the Swiss Lake Dwellings. The main sources used by the Chinese were in Sinkiang (East Turkestan), in the Kuenlun Ranges on the Tibet border and in Manchuria. Nephrite occurs *in situ* between hornblende schist and gneiss, but it was mainly obtained from boulders in river gravels. Nephrite was also used by the Maoris of New Zealand. It occurs as pebbles and boulders in the river valleys of north Westland in South Island and has been found *in situ* on Mount Cook and elsewhere in the Southern Alps.

Hornblende (Nos. 112-3), the common amphibole of many acid and intermediate and some basic igneous rocks and also of many metamorphic rocks derived from igneous rocks, is a complex calcium-magnesium-iron and sodium silicate. It occurs as prismatic Monoclinic crystals, black or greenish black in colour. The chief ways of distinguishing hornblende from augite are given on p. 136.

Crocidolite is an alkali amphibole, indigo-blue in colour and fibrous in structure. When infiltrated by silica it becomes a golden-reddish brown colour and is used for ornaments under the name of Tiger's Eye. (No. 38). The best material comes from Griquatown, South Africa.

Asbestos (Nos. 117-8). Mineralogically the term asbestos should be restricted to the fibrous forms of actinolite, but commercially the term has been extended to include fibrous anthophyllite, crocidolite, and chryosotile, a fibrous serpentine (p. 143). These minerals all occur as long fibrous crystals, which can be spun to form heat resisting (infusible) material. In addition, 'asbestos' has high electrical resistance and is immune to chemical action and decay. Chrysotile provides the bulk of the supply of asbestos. Its fibres are short, but extremely fine and strong. 32,000 feet of thread can be spun from one pound of chryosotile! The two main producing areas are at Thetford, Quebec and in the central Urals, where it occurs as veins traversing serpentinite. Anthophyllite, mainly mined in South Africa, provides much longer fibres, but their spinning qualities are markedly inferior, so that it is more often made into blocks. Tremolite may provide fibres up to a yard in length, but they are very brittle and so are more often used in wall insulation, whilst fibres of actinolite are commonly ground and mixed with the other varieties of asbestos.

The **Mica Family** are characterized by their perfect cleavage, which allows them to be split into thin elastic plates showing a splendent pearly lustre. The micas are important constituents of the more acid igneous rocks, of many metamorphic rocks and also occur derived in sediments.

Muscovite (common mica, white mica, potash mica, $2H_2O.K_2O.3Al_2O_3.6SiO_2$, No. 135 and Table p. 115) occurs as six-sided tabular crystals, which are Monoclinic, but appear to be pseudo-hexagonal. They may be white in colour or be tinged with shades of black, brown, green or yellow. Muscovite in economic quantities is obtained from granite pegmatites in which the micas occur in 'books' several feet across. The most productive mines are in India, near Biparad, Madras, but other deposits occur in New Hampshire, and North Carolina, U.S.A. The name Muscovite is derived from its use in Muscovy (Russia) as an early form of window glass.

Lepidolite (lithia mica, Nos. 136-7 and Table p. 115) usually occurs in small scales or granules. Its rose-red lilac or violet hue is very distinctive. It occurs in pegmatites, associated with such pneumatolytic minerals as tourmaline and topaz, in Madagascar, Elba, etc.

Biotite (dark mica, No. 148-9 and Table p. 115) is a complex silicate of magnesium, aluminium, potassium and hydrogen together with iron. It is the common variety of mica found in a wide range of igneous and metamorphic rocks. Its black or dark green colour is very distinctive, whilst thin laminae appear brown-green or blood-red by transmitted light.

Fuchsite (chrome mica, Nos. 140-1 and Table p. 116) is much less common than the other micas. Its strong green colour is very distinctive.

Phlogopite (magnesium mica) is typically of a brown bronzy colour. In Ontario it is worked where crystalline limestone has been invaded by an uncommon ultrabasic igneous rock, pyroxenite.

The micas, particularly muscovite, phlogopite and to a less extent biotite, have many important uses. They can be separated into sheets less than a thousandth of an inch in thickness and these are flexible, highly elastic, transparent, of low thermal conductivity and of high dielectric strength. They have many uses, particularly in the electrical industry.

Scapolite (No. 92 and Table p. 113) is a rare mineral found either in metamorphic rocks or as an alteration-product of lime-rich plagioclase feldspars in igneous rocks. It is a complex silicate of sodium, calcium and aluminium with some sodium chloride and calcium carbonate. When crystalline, it occurs as white or pale coloured Tetragonal crystals with a pearly to rather resinous lustre.

Garnet (Nos 93-6 and Table p. 113) is the name commonly given to a family of minerals, whose principal members are:-

Grossularite	$3CaO.Al_2O_3.3SiO_2$
Pyrope	$3MgO.Al_2O_3.3SiO_2$
Almandine	$3FeO.Al_2O_3.3SiO_2$
Spessartite	$3MnO.Al_2O_3.3SiO_2$
Andradite	
(Melanite)	$3CaO.Fe_2O_3.3SiO_2$
Uvarovite	$3CaO.Cr_2O_3.5SiO_2$

All garnets are Cubic, commonly occuring as rhombic dodecahedra (Fig. 11 IV) or as icositetrahedra (Fig. 11, VII) or as a combination of these (Fig. 24). Their hardness varies from 6.5 to 7.5, so garnets are used as abrasives, whilst some varieties are minor gemstones.

Grossularite, the common lime garnet, is typical of metamorphosed impure limestones. Its greenish colour is distinctive, whilst a yellower variety is known as cinnamon-stone.

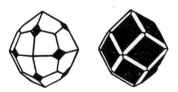

Fig. 24. *Crystals of garnet with rhombic dodecahedron and icositetrahedron in combination. On the left the dodecahedron (black faces) is suppressed, whilst on the right it is dominant.*

Pyrope occurs in ultrabasic igneous rocks and also in detrital deposits. In the 'blue ground' of the Kimberley mines (p. 119), it is found associated with diamonds. Of deep crimson hue, it forms very attractive gemstones.

Almandine is found in many metamorphic rocks. If brownish-red, transparent or opaque, it forms common garnet, if deep-red and transparent, it is precious garnet.

Spessartite is a rare variety occurring occasionally in acid igneous or in some metamorphic rocks. It is typically of a deep hyacinth colour.

Andradite is usually referred to as common garnet, but there is one variety, bright green in colour, found in the Ural Mountains, known as demantoid, which is the most valuable of the gem garnets. Melanite is a dull black variety, but most andradites are dark brown in colour. They occur in a wide range of igneous and metamorphic rocks.

Uvarovite, found in the Urals and in the Shetland Islands, in chromite-bearing serpentinites, is emerald-green in colour. It does not occur in pieces large enough to be cut for jewellery.

Olivine, $(Mg, Fe)_2SiO_4$, No. 97 and Table p. 113) is an important mineral in basic and particularly ultrabasic igneous rocks. One variety, forsterite, is formed by the metamorphism of an impure dolomitic limestone. When crystallized, olivine forms prismatic Orthorhombic crystals, usually in shades of green. They show a conchoidal fracture. Transparent pale green olivine, of gem quality, is known as *peridot*. It is found in Egypt, Upper Burma, Minas Gerais, Brazil. Unfortunately its hardness is only 6, so it is easily scratched when worn in rings.

Rhodonite, (Manganese Spar, $MnSiO_3$, No. 111 and Table p. 113) is sometimes used as an ornamental stone. Certainly the rose pink variety looks quite attractive when cut and polished. It occurs locally as a veinstone in lead and silver-lead ores.

Vesuvianite (Idocrase, a rather complex basic silicate of calcium and aluminium, No. 98 and Table p. 113) is a product of the contact metamorphism of impure limestones. Limestone blocks from Vesuvius yield perfect Tetragonal crystals (Fig. 11, VII), black or brown in colour.

Zircon ($ZrSiO_4$, No. 99 and Table p. 113) is an accessory mineral of igneous rocks, especially the more acid ones. It is also found in some metamorphic rocks. Owing to its hardness (7.5) it survives transport well and is a common 'heavy mineral' (p. 157) in arenaceous sedimentary rocks. Locally concentrations of zircon occur as in the gem gravels of Ceylon and in beach sands at Travancore, India, in Florida, Brazil, Byron Bay, New South Wales, etc.

Zircon occurs as prismatic Tetragonal crystals with adamantine lustre. *Zirconite* is grey or brown in colour whilst *Hyacinth* is a gem variety, red in colour and transparent. Another gem variety, colourless or smoky, is called *Jargoon*. The blue zircons which are another popular gem variety, are not natural, but are obtained by heat treatment of hyacinth. Unfortunately these stones are liable to discolour with time.

Zirconia, the oxide of zirconium, is extremely refractory withstanding temperature as high as 2300°C. and therefore is of importance in the chemical and electrical industry. The main source of zircon for these purposes are the zircon-rutile-ilmenite black beach sands of New South Wales, Travancore etc.

Topaz ($Al_2F_2SiO_4$, No. 100 and 1 and Table p. 113) is another accessory mineral of acid igneous rocks, especially those which have been affected by pneumatolysis (p. 154). It occurs in tin-bearing pegmatites and veins. It crystallizes as Orthorhombic crystals with a good basal cleavage. It is typically yellow, the pink colour of some jeweller's topaz being produced by the application of heat. It is an attractive gemstone. In the old days, all yellow gemstones used to be called topaz, but this omnibus term included citrine (yellow quartz, p. 124) and yellow corundum (p. 126). Topaz crystals of gem quality come from the Ural Mountains, Minas Gerais, Brazil, Ceylon and Saxony. In the British Isles beautiful sky-blue topazes have been found as water worn crystals in the Cairngorm mountains of Scotland and as colourless or pale crystals lining cavities in the granite of the Mourne Mountains, Ulster.

Andalusite (Al_2SiO_5, No. 102 and Table p. 113) is a significant metamorphic mineral. It is formed from argillaceous rocks under conditions of high temperature and low stress, occuring both in contact aureoles and in areas of regional metamorphism. Around Banff on the north coast of Aberdeenshire, a succession of interbedded argillaceous and arenaceous rocks have been metamorphosed. Those originally of clay grade now carry conspicuous crystals of andalusite, whilst the arenaceous rocks have been little altered.

Andalusite crystals are Orthorhombic, almost square in cross-section, grey or reddish in colour and are often altered on the outside to silvery mica. *Chiastolite* is a distinctive variety of andalusite, showing a cruciform cross-section (Fig. 25) with regularly arranged carbonaceous inclusions. Typical chiastolite-slates may be found in the contact aureole of the Skiddaw granite,

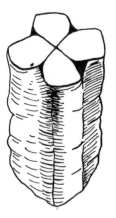

Fig. 25. *Chiastolite showing cruciform cross-section.*

Cumberland and in the aureoles of many of the granites of south-west England.

Sillimanite (Al_2SiO_5, No. 103 and Table p. 114) is another significant metamorphic mineral, indicative of higher stress conditions than those which produced andalusite from originally argillaceous rocks. It occurs as long needle-shaped Orthorhombic crystals and in aggregates, coloured shades of brown, grey or green.

Kyanite (Disthene, Al_2SiO_5, No. 104 and Table p. 114) is another significant metamorphic mineral formed from argillaceous rocks under conditions of high stress and moderate temperature. It usually occurs as long thin Triclinic crystals of attractive shades of blue, though sometimes the margins are colourless. In one respect it is unique amongst minerals; for its hardness varies significantly in different directions. The long bladed crystals can be scratched by a steel knife in the direction of their length (H. 4) but not across this length (H. 7).

Epidote (complex basic silicate of calcium, aluminium and iron, Nos. 105-6 and Table p. 114), is a common metamorphic mineral formed either from the alteration of impure calcareous rocks or of igneous rocks rich in lime-feldspar.

It is often granular, but if crystalline, occurs as flattened Monoclinic crystals, with perfect basal cleavage. Usually epidote is green in colour, but one variety found in Norway (No. 106) is almost black, but appears a dark oil-green on broken surfaces.

Other members of the Epidote family are *zoisite,* without iron and *clinozoisite* with some iron. As well as occurring in metamorphic rocks, they are formed from the saussuritization of feldspar (p. 134).

Beryl ($Be_3Al_3(SiO_3)_6$, No. 120-4 and Table p. 114) occurs as an accessory mineral in acid igneous rocks and in some metamorphic rocks. Beryl crystallizes in the Hexagonal system, forming six-sided prisms terminated by pyramid and basal pinacoid. Crystals of very large size, up to 18 feet in length and 18 tons in weight have been found. Ordinary Beryl is opaque and coloured shades of green, blue, yellow or white. It can, however, form magnificent gemstones; *emerald* in shades of green, (the colour being due to small amounts of chromium) and *aquamarine* in pale blue.

The finest emeralds come from near Bogota, Colombia, South America, where they occur in calcite veins, cutting bituminous shales of Cretaceous age; other important localities are in Brazil and the Ural Mountains. Many fine aquamarines have been found at Minas Gerais, Brazil, Madagascar and in California.

Tourmaline (No. 125-8 and Table p. 115) is a complex silicate of boron and aluminium together with varying amounts of alkalis (sodium, potassium, or lithium), magnesium or iron. It occurs as an accessory mineral in many acid igneous rocks, especially pegmatites and in some metamorphic rocks. It is usually formed as a result of pneumatolytic changes. Owing to the hardness it is a common 'heavy mineral' (p. 157) in arenaceous sedimentary rocks.

Tourmaline occurs as three-sided Trigonal crystals (Fig. 12, VII), which are unusual in being terminated by different faces at the two ends. Radiating groups of small needle-like crystals are also quite common. Its colour range

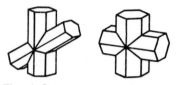

Fig. 26. *Interpenetrant twins of staurolite.*

is remarkable, whilst the colours are often arranged in zones parallel to the edges of a crystal. If a transparent crystal of tourmaline is held up to the light and rotated, it will be seen to be dichroic, that is it will change colour. A considerable number of varieties of tourmaline are recognized. Opaque iron-rich black crystals (No. 127) are known as *Schorl*. They are common in the pneumatolitically altered granites and 'country rock' of Cornwall and in places, as at the Roche Rock, a rock entirely composed of quartz and schorl can be found. *Rubellite* (No. 128) is a transparent red or pink lithia-rich variety, which is prized as a gemstone. The best gem quality tourmaline is found in pegmatite dykes, in California, in Madagascar, and in south-west Africa. Other varieties of tourmaline may show shades of blue, green or yellow.

Cordierite ($Mg,Fe)_2Al_4Si_5O_{18}$ with hydroxl, No. 129 and 130 and Table p. 115) occurs in high-grade regionally metamorphosed rocks and in contact-altered rocks. It is also found in some igneous rocks, owing to the assimilation of argillaceous material.

Usually granular or massive, when crystallized it forms short pseudo-hexagonal Orthorhombic crystals in various shades of blue. Some specimens from the gem gravels of Ceylon have been cut and used as a minor gemstone.

Staurolite ($OH_2FeAl_4Si_2O_{10}$, No. 131 and Table p. 115) is a metamorphic mineral formed from the alteration of argillaceous rocks. It occurs as Orthorhombic crystals, which are often twinned(Fig. 26). Colours are various shades of brown, whilst the surface is dull and rough.

The **Zeolites** are secondary minerals, formed from the alteration of feldspars and other aluminous minerals. They typically occur infilling cavities and amygdales, lining joints and other cracks in basalts and lavas. They are rather soft (H 3.5 to 5.5) and light in weight (S.G. 2 or slightly above). Only three of the large range of zeolites are considered here.

Apophyllite ($K_2Ca_4Si_6O_{12}(OH)_{16}$, No. 132 and Table p. 115) is Tetragonal with perfect basal cleavage. It is usually whitish or greyish in colour.

Natrolite ($Na_2Al_2Si_8O_8(OH)_2$, No. 133 and Table p. 115) is Orthorhombic, though more commonly it occurs as bundles of slender acicular crystals. Like so many of the other zeolites it is white in colour.

Heulandite ($CaO.Al_2O_3.6SiO_2.5H_2O$, No. 134) is Monoclinic and is white or brown in colour, it forms coffin shaped crystals.

Chlorite ($4H_2O.5(Mg,Fe)O.Al_2O_3.$ $3SiO_2$, No. 142 and Table p. 116) occurs either as a secondary mineral in igneous rocks, due to the alteration of biotite and other ferro-magnesian minerals, or in lowgrade metamorphic rocks – chlorite – phyllites, chlorite-schists, etc.

It is Monoclinic and may occur as tabular crystals, but more commonly in granular masses or as disseminated scales or as radiating aggregates. Its colour, various shades of green, is distinctive. It has a perfect basal cleavage and often occurs as flakes, which unlike those of mica (p. 130) are flexible, but not elastic. Further aids to identification are its rather greasy feel and its softness (H. 1.5-2.5).

Talc ($3MgO.4SiO_2.H_2O$, No. 145 and Table p. 138) is a secondary mineral, formed from the hydration of magnesium-bearing basic and ultrabasic igneous rocks and of dolomitic limestones. The changes may be due to contact-action of granitic bodies, stress during regional metamorphism producing talc-schists, or the action of magmatic waters. It normally occurs either massive or in flakes. Like chlorite the flakes are flexible, but not elastic. The diagnostic properties of talc are its greasy feel and its extreme softness (H 1). Its colour may vary from white through various shades of green.

Soapstone, Steatite is a massive variety of talc, which for long has been used for ornaments, especially by the Chinese. Talc has a variety of uses, ranging from the tailor's French Chalk, through fillers for paints and paper to lubricants and absorbents.

Serpentine ($3MgO.2SiO_2.2H_2O.$, Nos 143-4 and Table p. 116) is formed by alteration of magnesia-rich rocks, especially those containing olivine and pyroxene. It does not occur crystalline, but in massive, granular, or fibrous form. In colour it is extremely varied, green, black, brown, yellow or red, often veined with white steatite (see above) and brecciated. It is frequently mottled, hence its name after the appearance of a snake's skin. Its hardness is 3-4 so it can easily be cut with a knife and used as an ornamental stone and for ornaments. Visitors to the Lizard, Cornwall, will have had ample opportunity to see the uses and variety of serpentine. Large masses of serpentine also occur in Banffshire, in the Shetland Isles, in Galway, Norway, United States, etc. One rather attractive variety, *Ophicalcite* with green serpentine in white calcite, is formed by the metamorphism of a siliceous dolomite. *Crysotile* the fibrous variety of serpentine, which provides the bulk of commercial asbestos has been mentioned under amphiboles (p. 137). *Meerschaum,* used for bowls of pipes, is another hydrated magnesium silicate occurring in serpentine bodies. The main deposits of commercial importance are in Asia Minor, Morocco and Spain.

Glauconite (a complex hydrated silicate of iron and potassium, No. 146 and Table p. 116), is found in certain sedimentary rocks. It normally occurs as dull green or greenish-black grains, which may be sufficiently numerous to colour the rock green. Glauconite is easily oxidized to various oxides and hydroxides of iron, so in the weathered state many of the so-called 'greensands' are shades of brown or yellow. Glauconite is forming today on the sea floor off California and elsewhere. It is generally believed to be derived from the breakdown, under conditions of slow sedimentation, of ferromagnesian minerals, particularly biotite.

Sphene (Titanite, $CaO.TiO_2.SiO_2$, No. 147 and Table p. 116) is an accessory mineral in acid igneous rocks and is particularly abundant in contact-altered rocks rich in lime. It commonly occurs as wedge-shaped Monoclinic crystals, brown, grey, yellow or black in colour and of a high lustre. *Leucoxene* is a variety formed by the alteration of ilmenite and other titaniferous minerals.

Uranium and thorium are the two naturally occuring radioactive elements. Their average concentration in the earth's crust is extremely low. For uranium it is four parts per million in acid igneous rocks and two parts per million in the more basic igneous and sedimentary rocks. Extremely locally, however, pegmatite dykes and mineral veins may carry appreciable quantities of radioactive minerals. Both thorium and uranium are not stable, like most other elements, but undergo spontaneous disintegration with the liberation of α and β particles leading to the formation of radium. By measuring the degree and extent of this disintegration, it is possible to determine the time that has elapsed since a radioactive mineral was emplaced. In this way radioactive minerals have given a new precision to geológical chronology and enable the duration of the Eras and Periods of the Stratigraphical Table (p. 90) to be estimated in millions of years. With the recent development of nuclear power, there has been a most widespread and intensive search for workable concentrations of radioactive minerals. Great use has been made of the Geiger counter, which is a very delicate means of registering the proximity of radioactive sources. The recognition and distinction of the various radioactive minerals is a matter for the expert.

Pitchblende and Uraninite (Nos. 150-1) consist of uranium dioxide (UO_2) together with varying amounts of thorium, zirconium, lead, helium, argon, etc. Uraninite is Cubic, whilst pitchblende occurs in the massive, botryoidal or granular forms. Well known localities for pitchblende are Joachimsthal, Czechoslovakia, the Ka-

tanga district of the Congo, Great Bear Lake, Canada. Pitchblende is the main ore of radium. It is black in colour, but alters easily to yellow secondary minerals such as uranophane.

Monazite (No. 152 and Table p. 116) is only feebly radioactive. It is a phosphate of the Rare-earth elements (cerium, lanthanum and yttrium) with thorium and silica forming either ThO_2 or $ThSiO_4$. It is an accessory mineral in acid igneous rocks and may occur in large size in pegmatites. Locally, as at Prado, Brazil, at Travancore, India and in Ceylon, it has been concentrated in beach-sands formed from the erosion of nearby monazite-bearing granite masses. These sands are worked for their content of thorium and the Rare-earths, which have a variety of specialized uses, especially in the electrical industry.

Monazite occurs massive and also as flattened Monoclinic crystals, pale yellow to reddish-brown in colour and of a resinous lustre.

Steenstrupine (No. 153) is an extremely rare, feebly radioactive mineral, found at Julianehaab in Greenland. It is a fluo-silicate of the Rare-earths and thorium.

Euxenite (Polycrase, No. 154-5) is a complex oxide of the Rare-earths, thorium, niobium, iron, calcium, etc. It has been found in pegmatites in many localities in Scandinavia, Madagascar, Brazil, the Congo and North America.

Gadolinite (o. 156) is a complex silicate of yttrium – iron – beryllium with some thorium and uranium. It has been recorded from granite pegmatites, in many parts of the world. Some of the largest masses, up to 500 kilogrammes in weight, have been found in Norway.

Orthite (Allanite, No. 157) is a complex hydrated silicate of calcium, cerium, thorium, aluminium, iron, manganese and magnesium. It has been recorded from many localities mainly in pegmatites, but rarely as an accessory mineral in gneisses, schists and other regionally metamorphosed rocks. It is a member of the epidote group of minerals.

Asphaltite (No. 158) is the name given to a radioactive hydrocarbon found in certain Swedish pegmatites. It also occurs in some Canadian pegmatites and has been named there Thucolite after Thorium, Uranium, Carbon, Oxygen-lite. This name is now preferred to asphaltite. Its high lustre is noteworthy.

IV. THE IDENTIFICATION
OF ROCKS

The first step is to decide whether the rock is igneous, metamorphic or sedimentary. Igneous rocks are either glassy or crystalline, though in the finer-grained types, a good hand lens may be necessary to see the crystals. If the rock contains fossils, it must be of sedimentary origin. If it is bedded, it is very probably a sediment, but pyroclastic rocks may show bedding, whilst certain igneous rocks especially some lavas, may show a flow structure or banding (see No. 171) which might be mistaken for bedding. Metamorphic rocks are also often banded, but in this case, the banding is not flatsided, as in the sediments, but in the form of lenses (compare No. 226 with No. 246 or No. 261). The banding of metamorphic rocks may also be strongly folded and contorted (see No. 260). Sedimentary rocks are also often wholly or in part composed of clastic fragments (see No. 223, and Nos. 224-5). The pyroclasts are partly clastic in nature (see No. 164) but their fragments very rarely show signs of rounding and are usually composed of igneous material. Igneous and metamorphic rocks, unless much decomposed by weathering, are usually hard and compact. Sediments are much more varied, ranging from sands or clays to well consolidated rocks.

Igneous Rocks

Igneous rocks are classified as shown in Table III, first on their texture and then in terms of their chemical composition as indicated by the minerals present. In fine-grained rocks, the individual minerals are too small to be seen with the unaided eye. In medium-textured rocks, the grains in the groundmass are visible, but are too small for precise identification. These textural terms apply to the nature of the groundmass, in which may occur larger crystals (phenocrysts) of first generation-minerals.

As will be seen from Table III, the heavier a rock, the more basic it is, therefore as a rough working rule, light coloured rocks are most likely to be acidic, rocks dark in hue either basic or ultrabasic.

The fine-grained rocks are mainly of extrusive origin, quickly cooled lava flows. The medium and coarse-grained rocks are of intrusive origin, having been cooled under a cover, perhaps a cover several miles in thickness, of other rocks. Granites, gabbros, etc., are often referred to as Plutonic rocks, deep seated rocks, and dolerites and others which often occur as sills as Hypabyssal rocks, but to use the term Plutonic, Hypabyssal and Extrusive for major classes of Igneous rocks is to make assumptions from the texture of the rock alone as to its mode of occurrence. This can only be done after a thorough field study of the relation of the mass of igneous rock to the surrounding strata. In the classification used above, no direct assumptions are made as to the form and occurrence of the rock body. It is classified solely on features which can be studied in hand-specimen and, if greater precision is needed in mineral identification, under the petrological microscope.

TABLE III

Classification of the Igneous Rocks

TEXTURE	ACID	INTERMEDIATE		BASIC	ULTRA BASIC
		Alkali feldspar (orthoclase) predominates	*Soda-Lime feldspar (plagioclase) predominates*		
Coarse-grained	Granite	Syenite	Diorite	Gabbro	Peridotite, serpentinite, etc.
Medium-grained	Microgranite	Microsyenite	Microdiorite	Dolerite	—
Fine-grained	Rhyolite, obsidian	Trachyte	Andesite	Basalt	—
Specific Gravity	2.4-2.7		c. 2.8	2.9	3.0 or above
Essential Mineral Composition	Free quartz, feldspar, usually orthoclase, and mica. In the more basic types, some augite or hornblende	Hornblende and feldspar the dominant minerals. In the more acid types some quartz and/or mica, in the more basic types, some augite and/or feldspathoids (leucite, etc.)		Augite and plagioclase feldspar dominant, usually with olivine	Little or no feldspar. Often almost monomineralic

Sedimentary Rocks

The first broad division is into Clastic rocks, made up of fragments of other rocks, and the non-Clastic rocks, which are either largely or completely of organic origin, or were deposited chemically. The clastic rocks can then be divided on grain size, into those containing many particles greater than 2 mm. in diameter, those composed largely of sand grains and those of clay grade with particles 0.1 mm. or less in diameter. The calcareous rocks or limestones can be formed in a wide variety of ways. They vary in texture from coarse-grained rocks, containing clastic fragments or large fossils, down to extremely fine-grained rocks such as chalk. They are often fossiliferous, but sometimes the fossils are too small to be seen, except under a microscope.

The Sedimentary Rocks can be classified as shown below

1. Clastic Rocks	Rudaceous Rocks (coarse-textured)	Boulder and pebble beds
	Arenaceous Rocks (medium-textured)	Sands and sandstones
	Argillaceous Rocks (fine-textured)	Clay, shale and mudstone
2. Non-Clastic Rocks	Evaporite Deposits	Beds of gypsum, rock salt, etc.
	Ferriferous Rocks	Sedimentary iron ores
	Siliceous Rocks	Flint and chert
	Phosphatic Deposits	Phosphate rock and phosphatic pebble beds
	Carbonaceous Deposits	Peat, lignite, coal, asphalt, etc.
3.	Calcareous Rocks	Limestone, chalk, etc.
4.	Residual Deposits	Clay with flints, laterite etc.

The dividing lines between these groups are entirely arbitrary and each group contains rock-types which grade into one another and into the members of other groups. A typical limestone is easily distinguishable from a typical sandstone, but one may often come upon a rock consisting of quartz grains in a calcareous groundmass. Such a rock is transitional between two groups and can equally well be described as a sandy limestone or as a calcareous sandstone.

In identifying a sediment, the first thing is to decide whether it is a clastic or non-clastic rock. If clastic, look at its grain-size. In the more compact clastic rocks, the quartz and other grains are set in a matrix, which may be ferruginous (brown or red in colour), calcareous (will effervesce with dilute acid) or siliceous (will not effervesce or scratch with a knife). One can then describe the rock as a calcareous-sandstone, a ferruginous-sandstone, etc. Owing to the great variation amongst sedimentary rocks, such adjectival terms should be used whenever possible. A small plastic bottle of dilute hydrochloric acid is of great help in identifying sediments, for the calcium carbonate content of the rock can be estimated from the amount of effervescence when acid is placed up on it. All limestones, except those with a high content of dolomite (p. 159), effervesce vigorously with cold dilute acid and can also be scratched with a knife. If the sediment is not clastic or strongly calcareous, then it will belong to one of the non-clastic groups, whose characteristics are described later under the appropriate headings.

Metamorphic Rocks

Some metamorphic rocks, such as slate, a metamorphosed clay rock, and marble, altered limestone, are very distinctive. The metamorphism of rather mixed sediments (sandy limestone, etc.) or of igneous rocks will produce a wide variety of metamorphic rocks, whose character will depend partly on chemical composition of the original rocks and partly on the changes to which it has been subjected. These changes may include the loss of certain chemical constituents or of the addition of others. The two main classes of metamorphic rocks thus produced are the schists (finely banded) or the gneiss (coarsely banded). These rocks can then be subdivided in terms of distinctive minerals present into such rock-types as mica-schist (No. 249) and garnetiferous-gneiss (No. 261). Other metamorphic rock-types are described later.

V. DESCRIPTION OF THE COMMONER ROCKS

IGNEOUS ROCKS — 1 VOLCANIC

Pumice (No. 159) is light, both in colour and weight. It is formed during submarine eruptions, when gas-charged lava is cooled so rapidly that a cellular mass is produced. Pumice is light enough to float on water. In 1962 a Survey vessel in the Southern Indian Ocean steamed to investigate a reported submarine eruption, but it had to withdraw owing to the danger of the floating masses of pumice being drawn into the intakes of the engines. After a submarine eruption, pieces of pumice may be carried great distances by the waves.

Volcanic glass is formed when lava is quickly chilled. *Obsidian* (No. 160) is acidic in composition, black in colour and with a distinctive conchoidal fracture. It can easily be trimmed to give a sharp cutting edge and in areas where it occurs was often used by primitive man for his implements. Well known localities for obsidian are in the Yellowstone Park, U.S.A., on Mount Hecla in Iceland and the Lipari Islands in the Mediterranean. *Pitchstone,* which contains more water than obsidian, is another volcanic glass, red or brown in hue. Under the microscope, it is seen to contain much more crystalline material than obsidian, sometimes phenocrysts of quartz and feldspar in feathery-like microlites, which are incipient crystals. Thin selvages of pitchstone may quite commonly be found along the base of acid lava flows or at both margins of acid dykes and sills. *Tachylyte* volcanic glass of basic com-

position, may occur at the margins of basic lava flows, sills and dykes. Pelé's Hair, golden brown fibres of volcanic glass formed of lava spray, is an unusual variant found in the Hawaiian Islands.

Lava is molten rock material, which has been extruded on the earth's surface. The rock-types produced by lava (rhyolite, basalt, etc.) will be described later. We are concerned here with the form of lava flows. Acid lava is viscous and flows sluggishly, basic lava is more mobile and may flow down the sides of the volcano at speeds as great as 50 m.p.h., though the rate of movement is usually considerably less than this. The upper surfaces of newly consolidated lava flows may be either blocky or ropy lava, or to use the Hawaiian terms, aa or pahoehoe lava. Blocky lava is composed of a jumble of jagged blocks looking like a tumbled mass of slag. Ropy lava (No. 161) on the other hand, is formed when the gases contained in the lava stream escape tranquilly and not in a sudden burst. Sometimes after the upper surface of a flow has congealed, the remaining liquid drains away to form a lava cave, with icicles of shining black glass hanging from its roof. Such lava caves are one of the sights of Iceland. *Pillow lava* (Fig. 27) is indicative of extrusion either beneath water or the flowing of a lava stream into water. The outer surface of the pillows are pitted with vesicles (Nos. 185-6) infilled with calcite or other secon-

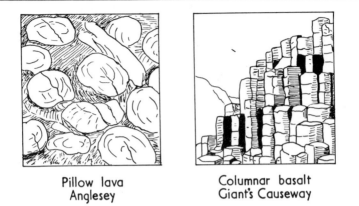

| Pillow lava | Columnar basalt |
| Anglesey | Giant's Causeway |

Fig. 27. *Pillow lava and columnar basalt. The pillows and the columns are each about 2 feet in diameter. Sediment between the pillows is indicated by fine shading.*

dary minerals. The concentrically arranged vesicles were formed by the gases escaping from the lava sphaeroid, which must have been jacketed for a short period with steam. The individual pillows may be separated by sedimentary material, or, if extrusion took place quickly, may be piled one upon the other. Spectacular pillow lavas may be seen on the south-west coast of Anglesey, near Ballantrae in Ayrshire and near Tintagel in Cornwall.

As the main mass of a thick lava flow cools, contraction occurs with the development of joints. With a regularly arranged and perfectly developed joint system, a columnar structure will be developed. Minor horizontal joints, often with a curved upper surface, cut the polygonal columns (Fig. 27), into blocks. The Giant's Causeway in Antrim, Fingal's Cave, Staffa off western Scotland are well known examples of the spectacular scenery thus formed. Columnar structures may also be produced during the cooling of intrusions. The long axis of the co-lumns is at right angles to the cooling surfaces, so in vertical dykes the columns will be horizontal, but in horizontally lying sills and lava flows, they will be vertical.

The **Pyroclasts,** the products of explosive volcanic activity, vary in grain size, from agglomerates (No. 164) which may contain blocks of rock several feet across to fine-grained tuffs which are consolidated volcanic ash. The larger fragments are usually angular, often comprise a variety of rock-types and are set in a fine-grained matrix composed of rock or mineral fragments.

Pyroclasts often contain lapilli or *volcanic bombs* (No. 163) composed of consolidated lava which had been blown high into the air. The spindle or sphaeroidal shape of the bombs is due to rapid rotation whilst in flight. Pyroclasts often show traces of bedding due to sorting by the explosion or to the material falling into water and settling according to its grain size.

151

Rhyolites are the Acid lavas containing free quartz which may be large enough in the porphyritic types to be recognizable by the unaided eye (No. 165 and 170). Typically these rocks are light in colour, relatively light in weight and often have a rather flinty appearance. They are often brecciated or flow banded (No. 171), a reflection of the extremely viscous nature of acid magma. Some rhyolites are sphaerulitic, (No. 280) the sphaerulites with their radially arranged fibrous crystals varying in size from a fraction of an inch to several inches in diameter. Owing to the high viscosity of acid magma, rhyolites are localized in their development. They are restricted to the immediate neighbourhood of the parent volcano.

Andesites and Trachytes, the Intermediate lavas, have more the character of rhyolite than of basalt in hand specimen (Nos. 182-3). They contain little or no free quartz, but if porphyritic enough, feldspar phenocrysts will be visible. In the trachytes the feldspar is dominantly orthoclase, in the andesites one of the plagioclase series. In such fine-grained rocks, even a hand lens may not give a great deal of information and for complete identification, examination of a thin-section under a petrological microscope may be necessary.

Andesites and trachytes are quite widely distributed, both as lava flows and as minor intrusions. In many volcanic areas, there is an interbedding of andesitic or trachytic lava flows with basaltic flows, the intermediate flows contrasting sharply with the more basic (basalt) flows owing to the much lighter colours which they show on weathering. Around the Pacific Ocean a well marked line – the Andesite Line –

can be traced. Outside the Line, in western North and South America, and the island arcs of Java, Sumatra, Japan and the Aleutian Islands, andesites are well developed, but within the Line, they are absent and the volcanic rocks of the central Pacific islands are all basaltic. The Andesite Line follows parts of the margin of the Pacific Plate, where oceanic crustal material is being carried downwards along inclined thrust planes. It has been suggested that friction along these planes has generated the heat that produced the andesitic volcanism.

Basalt is one of the most widespread and best known of rock-types. It is the main constituent of the thousands of square miles of plateau basalts in the Deccan in India, in the Snake River Plains in the U.S.A., in the Parana Basin in South America and in the now sundered Thulean lava fields stretching from Greenland, Iceland and Spitzbergen through the Faeroe Isles to Antrim and western Scotland. Extrusion of basaltic material, mainly as pillow lava, has occurred and is occurring along the Mid-Atlantic and other mid-oceanic ridges, whilst many basaltic volcanoes have been built up on the surface of the oceanic plates, particularly the Pacific Plate. The majority of these volcanoes are submerged, but the larger ones form the Hawaiian and other islands. Mauna Loa is really the highest mountain on the earth's surface, rising more than 30,000 feet above the surrounding sea floor. Basic magma is much more fluid than acid magma, so the gently sloping sides of the basaltic volcanoes contrast sharply with the more shapely andesitic volcanoes such as Fujiyama in Japan and the vertically sided spine that appeared after the eruption of the rhyolitic volcano of Mount Pelée in Martinique in the West Indies.

Outcrops of basalt often show *Spheroidal | or Onion Skin Weathering*. The corners of the joint bounded blocks (Fig. 27) have been attacked, producing weathered crusts round a central core, which will, in time, disappear.

Typically basalt is a dark, heavy, fine-textured rock (No. 184), but in detail there is considerable variation. Texturally it varies from basalt-glass to quite porphyritic types; mineralogically whilst basalt consists essentially of plagioclase and pyroxene, some varieties may contain much olivine, whilst, on the other extreme some may even show free quartz and mica. Many basalts are strongly vesicular (Nos. 185-6).

IGNEOUS ROCKS — 3 COARSE-GRAINED AND MEDIUM-GRAINED

Granite is another well known rock name. To the geologist, a granite is a coarse textured acid igneous rock containing visible quartz, feldspar and coloured minerals (usually mica). But to the non-geologist the term granite implies a hard durable rock, which is often of considerable commercial value. Hence the term 'granite' has been used in trade circles to include a considerable number of rocks, which are not of igneous origin, for example the 'Ingleton granite' of Yorkshire, a greywacke, and the 'petit granit' of Belgium, a limestone. We are here using granite only in its geological sense.

The coarse texture of granites is the result of the slow cooling of molten rock material at a considerable depth beneath the earth's crust. Granitic rocks normally occur in batholiths or smaller discordant masses in orogenic belts. They have been exposed by the erosion of the mountain chains. The granites of the younger mountain chains, such as the Dartmoor and associated granites of south west England, cut sharply across the structure of the 'country' rocks and are surrounded by a metamorphic aureole. Such granites must have consolidated at a high level in the earth's crust and are post-tectonic in the sense that they were emplaced after the folding of the country rocks. In the shield areas and in the older mountain chains, granites of a different kind may be exposed as a result of much deeper erosion. These granites occur in regionally metamorphosed rocks, but instead of cutting sharply across them, the granite is margined by a zone of migmatites (No. 262), in which the granite is intricately mixed and interlaminated with the metamorphic rocks. Such granites must have been emplaced before or during the folding and metamorphism that has affected the area. These pre- or syn-tectonic granites are very often very difficult to distinguish in hand-specimen from granite-gneiss (No. 243), which will be considered under metamorphic rocks.

Granites vary greatly in texture from strongly porphyritic types (Nos. 172, 203-4) such as the 'giant granite' of Cornwall with plenocrysts of white feldspar 6 inches or more in length, through a range of types of more uniform size (Nos. 194-202) down to the microgranites (Nos. 191-3) in which the different minerals are just recognizable with the unaided eye. Mineralogically they can be divided into the alkali granites and microgranites with orthoclase as the dominant feldspar, the adamellites with approximately equal amounts of orthoclase and plagioclase and the granodiorites with plagioclase predominating. Granites form attractive and durable decorative stones, as can easily be seen during a few minutes walk through the centres of most cities.

In addition to the normal granites, there are a number of special types, such as the spectacular *orbicular granites* (Nos. 210 and 211) with the rhythmically banded orbs composed mainly of coloured minerals set in a fine-grained mosaic of feldspar and some biotite. The *rapakivi granites* (Nos. 173 and 206) contain large rounded flesh-coloured crystals of orthoclase mantled with a narrow zone of white plagioclase and set in a matrix of quartz and coloured minerals.

During the final stages of the consolidation of granite magma, a quartz-rich solution, rich in volatiles is left and this is forced into cracks of the already frozen granite or into the country rock to form veins of pegmatite or aplite. *Pegmatites* are extremely coarse-grained and often contain extremely large and well-formed crystals (Nos. 214-5) which may be of the Rare-earth minerals. Beryl crystals found in a pegmatite in the Black Hills, South Dakota were up to 19 feet in length. Pegmatites often show *graphic texture* (No. 213) with the 'hieroglyphics' formed by small elongated prisms of quartz set in white microcline feldspar. *Aplites* are even-grained quartz-feldspar rocks with very minor amounts of other minerals (No. 212). Their texture is often referred to as saccharoidal, owing to a fancied resemblance to that of sugar.

The Cornish granites display a

variety of *pneumatolytic* modifications due to the attack of gaseous emanations following the final consolidation of the magma. Fluorine-rich gases produce *greisening,* leading to the formation of a rock rich in quartz, white (lithia-rich) mica and such fluorine-bearing minerals as topaz and fluorite. Boron-rich vapours produce first luxullianite, in which the feldspar has been largely replaced by dark tourmaline, and finally a *quartz-schorl* rock (No. 127), well seen at the Roche Rock, Cornwall. In *kaolinization,* the feldspars are attacked to form aggregates of flaky kaolin. In china stone, the half way stage, orthoclase is largely replaced and may also carry fluorite, whilst the plagioclase, albite, is unaffected. In the final stage, which produces china clay of such value to the ceramic industry, only the quartz is left unchanged in a friable mass of white feldspar, converted to kaolin.

Lamprophyres are a distinctive group of rocks, whose exact mode of origin is debateable. They occur as dykes, and in hand specimen show large phenocrysts, usually of mica, but sometimes also of augite, hornblende or olivine, set in a fine-grained groundmass. They are of very restricted occurrence. Alnöite (No. 179) shows the distinctive texture, but is abnormal in mineralogical composition, for as well as phenocrysts of biotite it contains much pink melilite, an uncommon complex silicate of calcium, aluminium and magnesium.

Syenites and microsyenites are the alkali-rich coarse and medium-textured igneous rocks. Mineralogically they differ from the granites by containing little or no visible quartz (No. 207) and by being essentially hornblende-feldspar rocks they tend to be both darker in hue (No. 205) and slightly heavier than granite. There are some very distinctive varieties of syenite, such as the Norwegian rock *Laurvikite* (Nos. 208-9) which is a widely used decorative stone. The bulk of the rock is composed of large feldspars, showing on polished surfaces a very handsome blue schillerization (No. 209). *Rhomb-porphyries* (Nos. 175-8) with their distinctly shaped feldspar phenocrysts set in a fine-grained groundmass are another group of well known Scandinavian syenites. Both laurvikite and rhombporphyries have been used as 'indicator erratics' for tracing the direction of movement of the Pleistocene ice sheets. Both rock-types outcrop in restricted areas of Norway and are also very easily recognizable. They occur in the boulder clays of East Anglia, proving that ice sheets from the Scandinavian mountains must once have extended right across the North Sea.

Diorites and microdiorites, the intermediate rocks with dominant plagioclase feldspar do not include types as easily recognizable in hand specimen as are some of the syenites. In hand specimen the diorites are usually even-textured rocks, composed essentially of plagioclase and hornblende. As the name suggests, in the dark-coloured hornblendites there is relatively little feldspar.

Gabbro, Norite and Troctolite, the coarse-grained basic igneous rocks, are typically dark coloured, fairly heavy rocks, made up of tightly interlocking crystals of feldspar and coloured mineral (Nos. 217-8). The feldspar is calcic plagioclase, usually labradorite and often showing schillerization. In the gabbros augite is the dominant coloured mineral, in the norites hypersthene or bronzite and in the troctolites olivine. Whilst many of these rocks have a normal granular texture, some show a distinctive *ophitic texture*, with large plates of augite enclosing plagioclase laths. This texture can sometimes be seen with a hand lens, but is most easily visible in thin-section. A typical troctolite is a striking rock with dark coloured olivines or pseudomorphs after olivine set in a background of grey or white feldspar (No. 219). In many basic complexes, in Skye, at Huntly near Aberdeen, at Freetown, Sierra Leone, in parts of the famous Bushveld Complex of South Africa and the Stillwater Complex of Montana, the basic rocks show a pronounced layering, varying in scale from layers hundreds of feet in thickness to half-inch bands of light coloured plagioclase-rich gabbro separating dark-coloured pyroxene-rich gabbro. The precise mechanism which produced this layering is still much under discussion and indeed it may well not be the same in every case. *Crystal sorting* or the rate at which the different minerals sank through the still molten magma is generally believed to have been an important factor. The speed of sinking of minerals would be controlled not only by their specific gravity, but also by their crystal habit.

Dolerites are medium-textured igneous rocks of gabbroic composition, normally occuring widespread as dykes or as sills, for example the Great Whin Sill of Northern England, along which part of the Roman Wall was built, and the numerous quartz-dolerite dykes of central Scotland. On the Continent and in America such rocks are often referred to as *diabases* (Nos. 187-190), but in Britain we prefer to restrict the term diabase to much altered dolerites. These rocks more often than gabbros, show ophitic texture, but this is usually only visible in thin section.

The **Ultrabasic igneous** rocks are typically coarse-grained, dark, very heavy and are often monomineralic. They are divided into the *peridotites* (No. 222) with olivine dominant and perhaps some other coloured mineral, but no feldspar, the *pyroxenites* with pyroxene dominant and the *picrites* in which some plagioclase is associated with olivine. Such ultrabasic rocks are usually found associated with gabbroic rocks in layered basic complexes.

Serpentinites (No. 281) are handsome rocks, often used for decorative purposes. They consist of the various serpentine minerals and are often streaked and blotched attractively in shades of green or red. They seem to be altered ultrabasic rocks, sometimes so completely altered that there is no indication of their original state, in other cases the serpentine seems to be pseudomorphic after pyroxene. Serpentinite may occur either in the basal parts of layered complexes, as at Stillwater, Montana, or as at the Lizard in Cornwall, in Anglesey, in the Shetland Islands on Unst, or in the Serpentinite Belts of New South Wales and New Zealand, as sheet-like bodies, that have been transported from considerable depths along great thrusts.

Eclogites are unusually heavy, striking looking rocks (No. 263) with bright red garnets set in a green matrix mainly of a soda-rich pyroxene, omphacite. They also contain small amounts of such unexpected minerals as quartz and kyanite and even diamond. Eclogites are very uncommon rocks. They occur in small outcrops amongst the oldest Pre-Cambrian rocks of the Scottish Highlands and Norway, whilst blocks are found in the diamond-bearing 'blue-ground' of Kimberley, South Africa. They are regarded as lying on the borderline between igneous and metamorphic rocks and as originating either as gabbroic igneous rocks which have been metamorphosed under extremely deep-seated conditions or as rock material which has been carried up, great thrusts from the base of the crust or the higher levels of the mantle.

Anorthosites are pure or nearly pure plagioclase rocks, consisting usually of labradorite. They may be quite light in colour (No. 220), but if the feldspar is schillerized then they are much darker in hue and more attractive looking (No. 221). On the weathered surface, small quantities of pyroxene enclosed in the feldspar stand out to give the rock a nodular appearance. Anorthosites occur as sheets in layered basic complexes or as great batholithic masses in the shield areas.

SEDIMENTARY ROCKS

Clastic Rocks

The **Rudaceous Rocks** are the boulder and pebble beds. They are composed of material worn off pre-existing rocks, transported a relatively short distance and then deposited in a rather unsorted condition. Rudaceous deposits are forming today on beaches, on the flood plains of fast flowing rivers, at the foot of rapidly wasting mountain slopes, etc. It is often possible to prove that fossil rudaceous rocks must have been formed in one of these situations. The fragments split off a rock face are usually angular, but as they are transported, their corners are rounded and eventually they became the perfectly smooth ellipsoidal shaped pebbles that can be seen on pebbly beaches. After they have been deposited, the loose boulder and pebble accumulations may be consolidated into compact rocks by diagenetic changes, such as the infilling of the spaces between the fragments by mineral matter deposited from percolating water. In this way boulder and pebble beds will be lithified into *breccias* (e.g. No. 282), which the majority of the fragments are angular or *conglomerates* with well rounded fragments predominating, (e.g. No. 283). One should also distinguish between a polygenetic rock (e.g. No. 224) with a wide variety of constituents and one that is monogenetic. The so-called 'Hertfordshire Puddingstone' of the London neighbourhood, (No. 283) made up almost entirely of flint pebbles in a siliceous matrix, is a good example of a monogenetic conglomerate. Sparagmite (No. 225) is a well known Scandinavian rock. In the specimen illustrated there are somewhat angular white fragments of quartz and pinker and smaller fragments of feldspar. Such a rock is rather on the border line between breccia and conglomerate and could be termed a *breccio-conglomerate*.

The material transported by a glacier is completely unsorted and varies in size from the finest 'rock flour' to masses of solid rock many yards in length. Where the glacier melts, its load is deposited to form *Boulder Clay*. As its name implies a most varied deposit, as can be seen at the moraines at the snout of any existing glacier or in the unconsolidated boulder clays that mantle so much of northern Europe and America. The existence of glacial conditions hundreds of millions of years ago is shown by the occasional

presence in the stratified rocks of beds of *tillites* or lithified boulder clay (No. 223). Such tillites occur in the Pre-Cambrian rocks of Norway, Canada, etc., and in the Permo-Carboniferous beds of South Africa, India, Australia, etc. Badly sorted polygenetic rudaceous rocks may be formed under a wide variety of conditions. The only certain proof that a suspected rock is really a tillite is for it to rest on a surface smoothed and striated by the passage of ice across it.

The Permo-Carboniferous tillites and other glacial deposits of India, South Africa, Australia, South America and Antarctica are believed to have been formed about 300 million years ago, when these now widely separated continents were joined together to form *Gondwanaland*. Since then Gondwanaland has broken up along plate boundaries and its fragments have been carried by Continental Drift into their present positions.

The **Arenaceous Rocks** are composed dominantly of material of sand grade. In general, they were deposited further from their place of origin than were the rudaceous rocks and so are both better sorted whilst the constituent sand grains are usually rounded to a greater or less degree. Arenaceous deposits are formed on beaches, in the shallower waters offshore, on the flood plains of rather sluggishly moving rivers, as coastal or desert sand dunes, as sand banks, etc. Diagenetic changes convert a sand into a *sandstone*. According to the type of cement one recognizes ferruginous-sandstone (No. 226), calcareous-sandstone, a siliceous or quartzitic sandstone, etc. There are many other variants, such as an *arkosic sandstone* (a finer version of No. 225) with grains of feldspar as well as of quartz, a '*greensand*' containing greenish or almost black grains of the mineral glauconite (No. 146), or a *greywacke*, an extremely illsorted sandstone with angular to subrounded grains of quartz, ferromagnesian minerals, etc., set in a groundmass of clay minerals, which have often been partly changed to greenish chlorite.

The great majority of sands and sandstones are made up almost entirely of grains of quartz. Owing to the prolonged sorting, other minerals have been disintegrated. Sometimes, however, one does find sandstones containing appreciable quantities of other minerals. Arkoses with abundant feldspar have been mentioned above. Gleaming flakes of mica are scattered along the bedding planes of 'micaceous sandstones'. Whilst a normal sandstone will contain very small amounts, less than a fraction of one percent, of '*heavy minerals*', (that is durable minerals such as rutile, tourmaline, zircon, iron-ore, etc., whose specific gravity is greater than that of quartz), under exceptional conditions of nearby exposure and subsequent concentration, arenaceous deposits may be formed rich in minerals other than quartz. Examples of these are the gem-bearing sands and gravels of Ceylon and Travancore, India, the black beach sands of New Zealand, rich in chromite and magnetite, the tin-bearing sands and gravels of Malaya with their high content of 'stream tin', etc. Such deposits are often of economic importance.

Sands and gravel deposits are worked very extensively as sources of aggregates for concrete and building sands. Deposits of exceptional purity, with a negligible iron content, provide *glass sands*. Another rather unusual deposit are *moulding sands*, for to be of good quality these require a 'bond' between the quartz grains enabling

Fig. 28. *A pebble of sandstone, showing true bedding in the upper and false bedding in the lower part.*

them to hold the shape required. Sandstones are widely used as *building stones*. If stained with iron compounds, they are both warm coloured and attractively variegated, as can be seen for example at the new Liverpool Cathedral, and also in the famous ruined abbeys of Tweeddale. The durability of the cement between the sand grains is clearly a prime factor in determining the life of a sandstone in a building.

Interest in arenaceous deposits, as in other sedimentary rocks, should not be restricted to the material of which they are composed. Much can be learnt from examination of the *Sedimentary Structures* shown on bedding planes and on surfaces at right angles to them. A sandstone may be well-bedded (No. 226) or show no sign of bedding (No. 228). It may show colour banding, due to slight variation in iron content (No. 227). Exposures or specimens of sands and sandstones often appear to be bedded in a variety of directions (Fig. 28). Careful inspection will usually enable one to distinguish between the *true bedding*, parallel to the original surface of deposition, and *false* or *current bedding* where the material has come to rest at the 'angle of repose' on the sides of sand dunes or subaqueous slopes. The true bedding is indicated by lines of small pebbles, larger sand grains, streaks of clay, etc., whilst the planes of false bedding are at an angle to this and are often curved. *Graded bedding* (Fig. 29) is due to the selective settling out of suspension of a rush of illsorted material carried across the sea or lake by a turbidity current. The sliding and movement of material down subaqueous slopes produces a wide variety of *Slump Structures* (Fig. 30). Greywackes often show both graded bedding and slumping.

On the bedding planes one should look for the trails and footprints made by organisms (Fig. 31), ripple marking (No. 229), suncracks (Fig. 32) and many other interesting and often problematical structures.

Argillaceous rocks, the clay rocks, are composed not of quartz grains, but of complex silicates, the *clay minerals,* which are the final product of the break down under weathering of pre-existing rocks. The clay minerals (montmorillonite, illite, kaolinite, etc.)

Fig. 30. *Section showing alternation of slumped and bedded layers.*

are of such minute size that the electron microscope, magnifying up to scores of thousands of times is needed for their study. The varying physical properties of different kinds of clay clearly depend on the kind and proportion of the clay minerals present, but much remains to be done in this field.

The clay rocks are deposited under quiet conditions, in the deeper parts of ponds, lakes and the oceans, but they may also be laid down on land, the finest products carried by wind action.

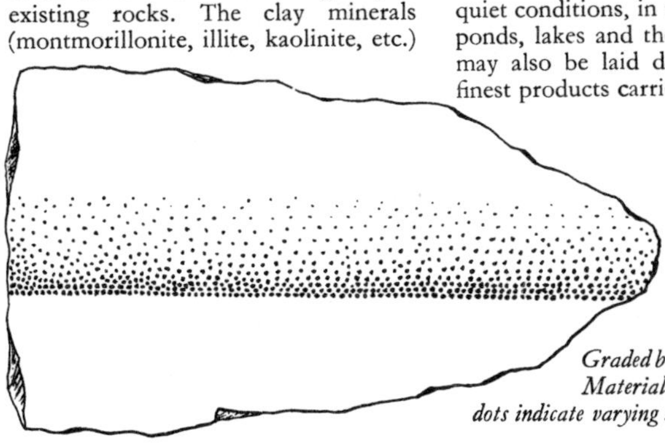

Fig. 29.
*Graded bedding in a greywacke.
Material of clay grade, white:
dots indicate varying size of sand particles.*

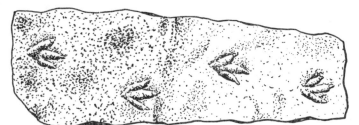

Fig. 31. *Footprints and trails.*
Above *Bedding plane with three-toed dinosaurian footprints*
Below *Bedding plane covered with numerous trails of unknown origin.*

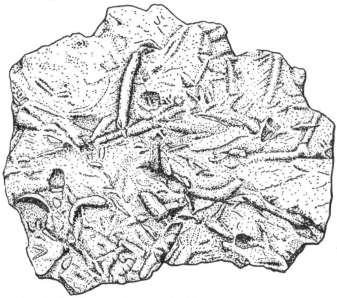

A *clay* as deposited under water has a high water content, but this will be gradually squeezed out by the weight of later strata and the clay harden into either a shale or a mudstone. A *shale* is a fissile clay rock with closely spaced bedding planes, whilst in a *mudstone* the bedding is much less well develop-ed. Many clays contain *concretions,* formed during diagenesis by the concentration of mineral matter. Concretions, when broken, may be structureless or may be *septarian nodules* (No. 289) showing a pattern of crudely radiating cracks infilled with calcite or some other mineral. Whilst clays have a

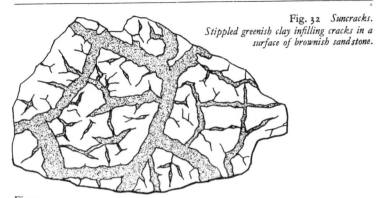

Fig. 32 *Suncracks.*
Stippled greenish clay infilling cracks in a
surface of brownish sandstone.

Fig. 33.
Examples of structures in which oil may accumulate. The porous reservoir rocks are stippled.

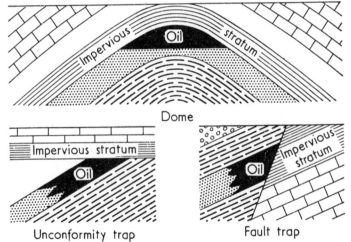

Dome

Unconformity trap

Fault trap

high water content, the extreme smallness of the pore spaces prevents easy movement of the water. One result of this is that whilst it is unusual for arenaceous deposits to contain fossils, for the calcareous bones and shells have been dissolved away by slightly acidic ground water able to

move freely through the large pores, clay rocks, on the other hand, are richly fossiliferous. The fossils are preserved either crushed flat along the bedding planes, or uncrushed in concretions, which have formed round them.

Clay rocks are smooth to the touch, a reflection of their extremely fine

161

grade. They are usually dark coloured (No. 230), in blues or greys if they contain much finely disseminated iron compounds, but becoming blacker with increase in carbon content. These are the colours of fresh unweathered clays, but when weathered they change to shade of red or brown, owing to oxidation of the iron compounds.

The main economic uses of clays are in the manufacture of bricks, as one of the raw materials for cement, and also to form watertight sealing material in engineering works. There are a number of special varieties of clay. *Pottery Clays* are white coloured, kaolinite-rich clays, laid down in ponds and lakes into which rivers have brought the products of the deep weathering of granite. *Fullers' Earth* is a peculiar argillaceous rock, which has many valuable economic properties. Some fullers' earths or bentonites seem to have been formed largely of extremely fine-grained material of pyroclastic origin, which has settled in water. *Alum Shale* (No. 213), is a particular variety of clay of Cambrian age, found in Scandinavia. As the name implies, it used to be worked for alum. Alum Shale contains much organic material and is quite strongly pyritic. The shale is interbedded with layers and concretions of 'stinkstone', which emit a foetid smell when hammered. Fossils are preserved in full relief in the stinkstone. Locally the Alum Shale contains asphaltite (p. 145) and is slightly radioactive.

Bone Beds can be regarded as a rare variant of the clastic rocks. They are thin beds containing a concentration of the teeth, bones, spines and other hard parts of vertebrates, usually fish. The vertebrate remains are easily recognizable owing to their high lustre (No. 284), and distinctive shape. Formerly

such bone beds were thought to be the result of some unspecified catastrophe, which had killed off the fish, etc., in great numbers. They are now, however, more commonly regarded as *condensed deposits,* laid down over a considerable period of time, when the vertebrate fragments were swept together into seams and pockets, and any finer-grained material settling on the lake or sea floor was carried away by bottom currents.

Non Clastic Rocks

The **Evaporite Deposits,** beds of gypsum, rock salt, etc. have been dealt with in the description of these minerals (pp. 128, 132).

The **Ferriferous rocks,** the sedimentary iron ores, are the different types of sedimentary rocks which contain a high enough content of iron for them to be economically important. They are really modifications of normal sedimentary rocks, laid down under conditions which produced abnormal concentration of iron compounds, either by chemical precipitation or by organic action or by a mixture of both. The bog iron ores of Sweden and elsewhere have already been mentioned under limonite (p. 127 No. 59). In England there are several different types of sedimentary iron ores. The Jurassic rocks which outcrop as a belt stretching from the Dorset coast between Lyme Regis and Weymouth to the East Yorkshire coast between Redcar and Scarborough, contain beds of limestones, often oolitic (see p. 167), with a high content of chalybite and siderite. The main ore fields occur around Banbury in Oxfordshire, in the Corby-Kettering neighbourhood of Northamptonshire, near Scunthorpe in Lincolnshire and in the Cleveland Hills of Yorkshire. The fresh ore is usually green in colour, but near the surface, and along joints, the iron carbonate minerals oxidize easily to brown limonite. It is believed that these ores were laid down in shallow current-agitated seas, in which conditions were exceptionally favourable for the chemical precipitation of iron compounds carried in solution by rivers from the neighbouring land areas. Such iron ores may contain fossils with the hollow interiors of the shells partly infilled by crystals of calcite growing inwards from the walls (No. 286) below. In the Middle Ages and even later, the Weald of Kent, Surrey and Sussex used to be the great iron field of England. Here the ores occur in shales not as continuous beds, but as lines of concretions of sideritic clay-ironstone, often oxidized externally to limonite. These ores are not of marine origin, but were laid down in swamps, where abundant decomposing vegetation helped the precipitation of the riverborne iron compounds. Locally in the Coal Measures, occur the 'Black-Band ores' consisting of alternations of coaly and ferruginous material. In South Wales there are local deposits of iron ore; in this case a limestone in which the pores of the rock and its contained fossils have been infilled with ferric oxide. The ores of the Clinton iron field in the Appalachians are of a similar nature.

In many of the Pre-Cambrian shield areas, e.g. the Lake Superior region in North America, in Brazil, in East Africa, sedimentary iron ores occur a different character. This ore is markedly banded with alternating layers of chert, often somewhat stained with haematite, and layers of iron compounds (hematite, goethite, etc.). These banded ores (No. 287) may be hundreds of feet in thickness. Their exact mode of formation is still vehemently debated.

In sands and sandstones one often finds layers and ramifying seams, often of the most bizarre appearance, of 'ironstone' or 'carstone'. Limonite has been deposited from percolating water to bind the sand grains together. Such seams have far too high a silica content to be worked as an iron ore and therefore should not be classed as ferriferous deposits.

Siliceous Rocks. Flint and chert have already been described on p. 124. *Diatomite* or diatomaceous earth occurs as small deposits in Germany and other continental areas, and is forming today on the floors of some Scottish lakes. It is extremely fine-grained, for it is composed of myriads of the small siliceous skeletons of diatoms. These deposits were laid down in small ponds and lakes, in whose waters diatoms flourished in great abundance. After death their skeletons sank to the bottom. If little or no other material was brought into the lakes, then a very even-textured deposit of extremely high silica content accumulated. Such deposits are of considerable economic importance and are used for many purposes, mainly insulating and soundproofing, in which the inert nature of the diatomite is an advantage.

Phosphatic Deposits are of two kinds; bedded deposits and beds of phosphatic nodules. In Algeria and parts of the Rocky Mountains, there are extensive beds of phosphate-rich limestones of marine origin. Locally also the Chalk of England and France is phosphate-rich. At quite a number of horizons in the stratified rocks, one finds beds of nodules of grey or black phosphatic material in which are embedded quartz grains and fossils. The nodules are often rounded, and sometimes have oysters growing on them or are bored by other organisms. Such beds must have been formed by the submarine erosion of bedded phosphate deposits, followed by the concentration of the nodules. Phosphate deposits are being formed today on certain tropical islands, such as Nauru in the Pacific, by the 'guano' from the droppings and skeletons of myriads of sea birds.

Carbonaceous Deposits. *Peat* is forming today, or has formed during the last few thousand years, in illdrained areas with abundant plant growth. Retardation of the normal breakdown of plant tissue by bacterial action produces a loose spongy brown material, full of easily recognizable plant debris. The extensive spreads of peat on many of the mountains of the Pennines, North Wales and Scotland show that during the 'Climatic Optimum' of about 4000 B.C., these areas had a thick vegetation cover. Layers containing the stems and trunks of forest trees can often be seen.

Conditions favourable for the accumulation of thick beds of peat have occurred many times during the geological past. As the peats were buried by later deposits, they were compressed and affected by slow chemical changes, especially the loss of their volatile constituents to be converted with the passage of time, first into Lignite (Brown Coal) and later into Humic Coal.

Extensive deposits of *Lignite,* formed a few tens of millions of years ago, occur, in many parts of Europe. In places, individual seams may be as much as 100 metres in thickness. Lignite (No. 238) is clearly compacted, but still contains recognizable stems and leaves of plants. *Humic Coal* is of much greater antiquity. The coal seams

of most parts of the world originated as peat beds several hundreds of millions of years ago. The peat has now been converted into a fairly hard black rock, which is usually banded with some layers lustrous, others appearing dull looking. Flattened impressions of leaves can sometimes be seen, but, in general, there is little obvious sign of vegetable origin. In the humic coals, one can recognize a series of increasing 'rank' from the house or coking coals through the bituminous coals, which soil the fingers, to the steam coals or anthracites which do not. This series reflects increase in content of carbon, decrease in volatiles and also, to some extent, increase in calorific value. Thin films and small nodules of pyrites are a feature of many coal seams.

Carbonaceous material may also accumulate on the bottoms of seas and other bodies of water in fine-grained sediments with a high organic content. If anaerobic conditions prevailed, that is there was a deficiency of oxygen, so that the normal processes of organic decay were retarded or even prevented, then by a complex series of changes not yet fully understood, the organic material in the sediments might be converted into petroliferous hydrocarbons or *mineral oil*. As later sediments were deposited, the mineral oil was very prone to migrate upwards or sideways from its place of formation into rocks with sufficient fissure or pore space to act as reservoirs. If further movement was prevented by an impermeable rock, then the oil would accumulate in one of the types of structure shown in Fig. 33. If the cap rock is punctured by a drill, an oil producing well will result. If, on the other hand, there is no cap rock or if the cap rock is worn away by later erosion, then the oil will gradually seep upwards to the surface of the ground.

As it evaporates, it may leave behind deposits of the heavier and more viscous fractions, such as the asphalt which slowly wells up into the Pitch Lake of Trinidad or the rubbery asphaltic deposits (No. 158) which occasionally line fissures in rocks to show that mineral oil has passed through them.

Amber (No. 149) is fossil resin, which sometimes contains insects that were trapped in the gum extruded from coniferous trees. It is found as pebbles in the Tertiary strata of the south Baltic coast and has long been used for beads, ornaments, etc.

The **Calcareous Deposits,** the *Limestones,* show a great range of variation. Owing to the comparative softness of calcite (3 on Moh's Scale as compared with 7 of quartz), fragments of limestone soon disintegrate during transportation. They may, however, travel for great distances if frozen in icesheets. Some boulder clays are very rich in

fragments of limestone. Calcareous breccias and conglomerates do occasionally form at the foot of wasting cliffs of limestone. Another distinctive property of calcite is its high degree of solubility in slightly acidic water. Limestone regions in many parts of the world are riddled by cave systems due to the underground solution of the limestone by water moving down joints and along bedding planes. Material falling from the roof, or washed into the mouths of these caves may form *cave breccias,* which may enclose the bones of creatures that lived in or near the caves. Famous examples of such cave breccias are those which have yielded the remains of the hominoid australopithecines at Sterkfontein in South Africa, those which contained the remains of 'Peking Man' or as at Kent's Cavern near Torquay in Devon, have enclosed human implements and the bones of the contemporary cave bear, hyena, etc. Some limestone caves are also famous for the paintings on their walls. Another well known

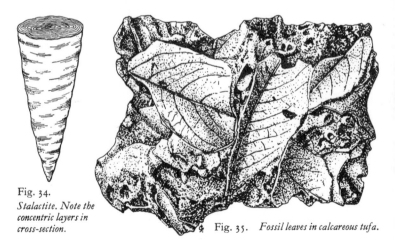

Fig. 34.
Stalactite. Note the concentric layers in cross-section.

Fig. 35.   *Fossil leaves in calcareous tufa.*

feature of many caves are the *stalactites,* hanging from the roofs and the *stalagmites* rising from the floor. These are formed from the evaporation of lime-charged water. They usually show a concentric structure (Fig. 34) and may be attractively stained by small traces of iron or copper salts. *Tufa* is formed round the so-called petrifying springs, whose water is oversaturated with calcium carbonate and this is deposited to enclose leaves, twigs, etc. (Fig. 35). *Travertine* is a more compact and banded variety, often worked as an attractive decorative stone. The polished slabs of travertine lining the halls of the Senate House, University of London, show an intricate pattern of dark and lighter coloured layers.

The calcium carbonate dissolved by rain falling on the land areas passes into solution as calcium bicarbonate and is carried by rivers into the lakes and oceans. There the bicarbonate may be chemically precipitated or it may be abstracted by a wide variety of organisms to build up their skeletons. Many

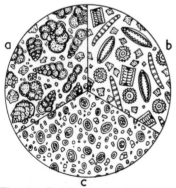

Fig. 36. *Rock forming microfossils. Separation of (a) foraminifera* × 40; *(b) diatoms* × 40; *(c) coccoliths and coccospheres* × 350.

different types of limestone can be recognized. Some limestones are built up very largely of organic material (Fig. 37). The modern 'coral reefs' are one example, though it must be stressed that corals form only a subordinate part of the framework of the reef and that the lime-secreting algae are much more important. In the stratified rocks, a variety of reef-limestones occur (e.g. No. 236), a number of them built up by other organisms than corals. The organic content of limestone is sometimes of extremely small size. The exact nature of the familiar *Chalk,* an extremely fine-grained rather soft white limestone of exceptional purity, could not be ascertained with certainty until the electron microscope had been invented. With its extremely high magnification it was possible to show that chalk is largely composed of the remains of coccoliths (extremely minute calcareous algae) together with the shell debris of other organisms. It is customary to name a limestone of organic origin after the main kind of organism present, e.g. the Orthoceras limestone of No. 232-3, the gastropod limestone of No. 285, the crinoidal limestone of Fig. 37 etc. As organisms accumulate on the sea floor, they may be worked over and broken up by bottom currents to form a *Bio-clastic* limestone with the broken and rolled organic fragments set in a fine-grained matrix, which may be, in part, a chemical precipitate. Calcium carbonate is unusual, for its solubility decreases with rise in temperature and therefore chemical precipitation is particularly likely to occur in hot rather shallow seas. Under these conditions and with fairly vigorous bottom current action, *Oolitic limestone* will be formed, made up of spherical or ellipsoidal ooliths, with a well developed concentric structure, set in a fine-grained ground-

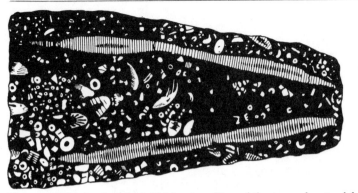

Fig. 37. *Polished surface of crinoidal limestone. Pieces of the stems and scattered fragments of the crinoids (sea lilies) are preserved in white calcite in a very fine-grained dark-coloured matrix.*

mass (Fig. 38). In a high-quality oolite, such as the well known Portland and Bath building stones, the ooliths are all very much of the same size, for this is determined by the maximum size of the particles which could be moved by the bottom currents. Not all of them can withstand the polluted atmosphere of cities and industrial towns. Bath Oolite weathers much more quickly on a London building than does Portland Oolite. Many limestones, particularly fossiliferous limestones, cut and polish very attractively. Such rocks are often given the trade name of Marble, for instance 'Purbeck Marble' (No. 285). But geologically this is a misnomer, for they are sedimentary and not metamorphic rocks. Vast quantities of limestone are quarried, to be mixed with clay and calcined in great rotating kilns to form cement. Hydraulic limestones are certain argillaceous limestones or thin beds of limestone separated by layers of clay or shale, containing calcium carbonate and aluminium silicate in the correct proportions to form a natural cement

of fossils and clearly are largely of organic origin, others show every sign of having been precipitated, others are clearly of mixed origin Many limestones contain layers or nodules of flint or chert.

Calcareous rocks may contain dolomite as well as calcite. A true dolomite rock is composed entirely of the

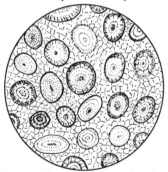

Fig. 38. *Thin-section of an oolitic limestone* × 20. *The circular to ellipsoidal ooliths are enclosed in a fine-grained matrix of crystalline calcite.*

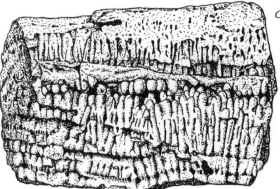

Fig. 39.
Concretionary dolomite;
Fullwell, Sunderland.

mineral dolomite, but these are much less common than the dolomitic limestones, containing both dolomite and calcite. Dolomitic rocks tend to be honey-coloured. The rhombs of dolomite may be visible with a hand lens, whilst a completely dolomitized rock will not effervesce with cold dilute hydrochloric acid. The calcite and the dolomite grains are usually not intimately mixed. The dolomite may occur as concretions in a calcareous rock, as in the Chalk of Faxe in Denmark (No. 237) or as in the Cannon Ball Limestone of Sunderland, spheres of calcite may be set in a matrix of less well consolidated dolomite. Sometimes these concretionary structures assume the most fantastic shapes (Fig. 39).

Dolomitic limestones may be either of primary or of secondary origin. The primary dolomites were chemically precipitated with the two carbonates coming out of solution together, whilst in the secondary types an original calcitic rock has been altered after deposition by percolating magnesium-charged solutions. Secondary dolomitization is often selective, unaltered fossils of calcite being preserved in a dolomitized matrix.

Many calcareous rocks form a good building or decorative stone, but not all of them can withstand the polluted atmosphere of cities and industrial towns. Bath Oolite weathers much more quickly on a London building than does Portland Oolite. Many limestones, particularly fossiliferous limestones, cut and polish very attractively. Such rocks are often given the trade name of Marble, for instance 'Purbeck Marble' (No. 285). But geologically this is a misnomer, for they are sedimentary and not metamorphic rocks. Vast quantities of limestone are quarried, to be mixed with clay and calcined in great rotating kilns to form cement. Hydraulic limestones are certain argillaceous limestones or thin beds of limestone separated by layers of clay or shale, containing calcium carbonate and aluminium silicate in the correct proportions to form a natural cement mixture. Another important use of limestones, including dolomite, is to serve as a flux in blast and other furnaces. Limestone is also extensively quarried for use as agricultural lime to reduce the acidity of soils, as a top dressing for roads and for many other purposes.

The **Residual Deposits** are the result of chemical weathering of other rocks. Soluble material is removed in solution and a crust of the insoluble components slowly accumulates on the land surface. The ridges of the chalk lands of much of southern England and northern France are capped by an irregular thickness of reddish *Clay with Flints*. The *terra rossa* of the limestone areas of Yugoslavia and southern France is of similar origin. These are the deposits formed in temperate latitudes. In the tropics, where both rainfall and temperature are higher, chemical weathering is more intense and not only calcareous rocks are affected. Over vast areas of tropical Africa, India, unaltered rocks are hidden beneath a continuous spread, up to a hundred feet in thickness, of *Laterite*, mottled in shades of red and brown. Laterite is soft when freshly dug, but quickly hardens on exposure to form good building material. It has resulted from the intense chemical weathering of a wide variety of rocks, basalt, gneiss, schist, slate, granite, etc. Much of their silica content has been removed in solution to leave behind the hydroxides of ferric iron, aluminium and some manganese. If the residual deposit is highly aluminous, with little or no ferric oxide, it becomes a *Bauxite* and may be worked as an ore of aluminium. At the Giant's Causeway, Antrim, the flows of dark-coloured basalt are interrupted by a conspicuous bed of red bauxitic laterite. This bed gives evidence of a pause in the volcanic activity long enough for deep weathering to have occurred at a time when the climate of Britain must have been tropical.

METAMORPHIC ROCKS

Metamorphic rocks differ from igneous and sedimentary rocks both in mineral compositon and in their structural features. These differences reflect the conditions of stress and high temperature under which pre-existing rocks were given a distinctive new character. We have already referred (p. 160) to the ways in which rocks may be changed by weathering or by percolating waters at shallow depth; but these are metasomatic, and not metamorphic, changes.

Metamorphic rocks commonly show indication of the very severe pressure to which they have been subjected. They may show *cataclastic structure*, such as the stretching out of pebbles in a conglomerate (No. 246) or *augen structure* (No. 243) with large crystals of pink feldspar, eye shaped or lenticular in form lying with their long axes parallel to each other, or on a smaller scale, they may show a *lineation* or preferred orientation of the minerals which imparts a definite 'grain', (running vertically in No. 240), to the rock. Coarser banded rocks may be obviously folded (No. 260), whilst rocks of the clay grade may have new planes of parting or *cleavage*, which are at an angle, often a high angle, to the original bedding.

Some common minerals, quartz, the feldspars and the micas, in particular, occur in metamorphic as well as in igneous and sedimentary rocks. But there are a number of minerals, which originate in and therefore are distinctive of metamorphic rocks. Garnet is the most typical, but we may also mention andalusite, cordierite, and staurolite. In many metamorphic rocks, these minerals may occur in crystals large enough to be recognizable with the unaided eye. They may also occur in sediments, but then only as worn grains derived from the erosion of metamorphic rocks; grains that can only be recognized under the microscope and that are present in extremely small amounts.

The nature of a metamorphic rock is due partly to the original composition of the rock which has been metamorphosed and partly to the type and extent of metamorphism which it has suffered. It therefore follows that similar or nearly similar end products may have had considerably different histories, – histories which may only be interpretable, after complete chemical analysis and very detailed microscopical investigation. In describing the main metamorphic rock-types, we shall indicate some of the ways by which each may have been produced, but it must be understood that there are other possibilities.

Quartzites (Nos. 247-8) are the result of the metamorphism of pure arenaceous rocks. As only quartz is present, there can be no development of new minerals, but the quartz grains may have recrystallized to form a tightly interlocking mosaic and hence a tough brittle rock, which does not show the clastic grains of quartzite of sedimentary origin. With an increasing amount of original impurities, metamorphism will produce quartzites which are streaked or mottled with secondary minerals (No. 247), until the quartzites grade into some of the other rock-types described below.

Leptites (Scandinavian term), *Granulites* preferred in English speaking countries (No. 239) are even-textured

tough rocks, showing a crude foliation due to the parallel alignment of flat-sided lenses of quartz and feldspar. The granulites of the Pre-Cambrian rocks of the Scottish Highlands contain thin dark micaceous layers, which suggest relic original bedding, but are really the result of foliation. Under the microscope many granulites are seen to contain grains of almandine garnet kyanite, sillimanite and other high temperature minerals. Granulites are clearly the product of high-grade regional metamorphism involving high temperatures and pressure.

Slates (No. 251), are produced by the metamorphism of argillaceous rocks. They are *pelitic* rocks of clay grade as distinct from quartzites and granulites, which are *psammitic* rocks of sand grade. Mineralogically slates are composed of white mica, green chlorite and quartz together with smaller quantities of other minerals. One of the most distinctive features of slates is their perfectly fissility or *cleavage,* which may be at any angle to the original bedding. In quarries in the Slate Belt

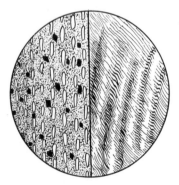

Fig. 40. *Thin-sections of slates* × 20. *Left - True cleavage; right - False cleavage.*

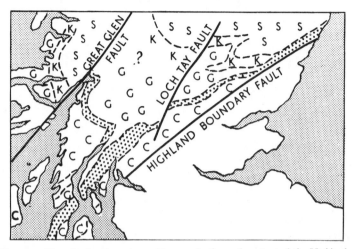

Fig. 41. *Map showing the metamorphis zones in the southern part of the Highlands of Scotland. S - Sillimanite zone; K. - Kyanite zone; G. - Garnet zone; Stippled - Biotite zone; C. - Chlorite zone. The rocks south of the Highland Boundary Fault are not metamorphosed.*

of North Wales one can easily find specimens showing the difference between cleavage and bedding. The shape of the specimen is clearly determined by the cleavage, whilst coarser-grained greenish bands indicate the bedding (No. 290). Cleavage may be of two kinds. In *true or slaty cleavage* the flakes of mica and chlorite have a perfect parallel alignment whilst the less perfect, *false, strain-slip or fracture cleavage* is due to closely spaced minute folds, along whose axes the chlorite and mica flakes have been sharply bent. (Fig. 40). Cleavage is the result of low-grade pressure or dynamic metamorphism, but in the vicinity of igneous aureoles, slates may become spotted due to the development of crystals of cordierite and other high temperature minerals.

Phyllites (No. 250) have the same mineralogical composition as slates, but are notably coarser. The micas, in particular, are large enough to give a characteristic silky sheen to the rock. Phyllites may show thin quartzo-feldspathic layers and thus grade into the schists. Phyllites are the result of rather more intense metamorphic change than that which produces slate. Perhaps higher temperature or longer duration of stress conditions.

Schists (No. 249) indicate the still greater metamorphism of pelitic rocks. They show a pronounced foliation or parallel alignment of the constituent minerals. Careful inspection of schists, particularly those from intensely deformed regions, will often reveal the presence of several directions of schistosity with the most pronounced (later) ones almost, but not quite, obliterating the earlier ones. Clearly the tectonic history of such a rock is a complex one.

Schists can be produced under a considerable range of metamorphic conditions and one can infer much from the mineralogical composition of the schists. Mica-schists, which are the end member of the series slate → phyllite→mica-schist, are the result of the low grade metamorphism of pelites. Quartzo-feldspathic schists with less mica originated as more psammitic rocks. An impure limestone under low grade metamorphism will give rise to calc-schists made up of coarse-grained calcite with quartz, muscovite, chlorite, etc. Green schists with an abundance of chlorite, epidote or actinolite are the result of the low temperature regional metamorphism of basic igneous rocks, whilst the talc-schists were originally peridotites.

Under conditions of greater temperature and pressure, – high grade metamorphism – new minerals notably the 'stress' minerals, garnet, staurolite, sillimanite and kyanite will be produced. Garnet-mica schist with its deep red almandines is a very characteristic and easily recognizable rock-type. In parts of the Scottish Highlands a regular zonation of the index minerals of the schists can be recognized (Fig. 41). The high grade calc-schists are characterized by the green lime-rich garnet grossularite and by diopside, a lime-rich amphibole and by epidotic minerals such as clinozoisite.

Hornfels (No. 252), in contrast to the strongly foliated rocks described above, is a tough compact flinty looking rock. It is composed of a mosaic of tightly interlocking mineral grains, all much of the same size and showing no preferred orientation. Many hornfelses are fine-textured and it is difficult to recognize individual minerals, even under a hand lens. The principal minerals are quartz, feldspars, pyroxenes, grossularite and calcite, together with larger crystals, (porphyroblasts) of andalusite and cordierite, indicative of thermal metamorphism. Hornfelses form the inner part of the contact aureole of many large plutonic masses and grade outwards into spotted slates and spotted schists. As they are the result simply of thermal metamorphism without deformation, they may show relic structures of their parent rocks, e.g. amygdaloidal textures in rocks that were originally volcanic, banding in altered sediments, though in this case there may be an interesting reversal of grain size. The original fine-grained argillaceous layers are now represented by bands rich in large flakes of mica and large grains of andalusite, whilst the originally more sandy layers have recrystallized, without any appreciable change of grain size, into finer-textured-quartzo-feldspathic bands.

Gneisses (Nos. 255-261) are coarse-grained irregularly banded rocks, with schistosity much less well developed than in the schists. Quartzo-feldspathic layers alternate with mica-rich layers. They are one product of regional metamorphism, especially of high grade metamorphism. The terms para- and ortho-gneisses are used to distinguish gneisses which originated as sediments from those which were of igneous origin. Many gneisses are, however, so altered that it is not possible to apply these terms with safety. The mineral composition of gneisses is similar to that of the high grade schists.

Injection-gneisses or granite-gneisses (No. 242) are rocks which have originated at depths and temperatures great enough for metamorphic rocks to be thoroughly permeated and injected by granitic liquid. The resultant rocks are of gneissose texture, but essentially of granitic composition.

Amphibolites (No. 240) are medium to coarse-grained rocks, composed essentially of hornblende and plagioclase. They commonly show a distinct lineation, due to the parallel alignment of the hornblende prisms. They are usually the product of the medium to high grade regional metamorphism of basic igneous rocks. It is sometimes possible, as in parts of north-west Scotland, to follow a dolerite dyke along its strike and to trace its change into an amphibolite as the grade of metamorphism of the country rocks becomes more intense.

Marbles (No. 254) are the result of the thermal or regional metamorphism of limestone. If the original rock was pure limestone, then as for quartzites (p. 162), the only change can be recrystallization, to produce a tightly interlocking mosaic of calcite crystals, as in the well known white Carrara Marble of Italy, so popular for use in statuary and in cemeteries. But if the parent rock was an impure limestone or a dolomitic limestone, then a much more attractive looking marble may result. Various lime-silicate minerals, such as wollastonite, forsterite, brucite, the lime garnets, etc., are formed to produce most pleasantly variegated rocks, which cut and polish easily to form excellent decorative stones. As already mentioned (p. 168) the term marble has a commercial value and many so-called marbles (e.g. No. 253) are not metamorphic rocks, for they contain recognizable fossils and are really compact unaltered limestones, which can be cut and polished.

Skarns are formed where igneous rocks, particularly acid ones, have been intruded into calcareous rocks. A narrow zone of metasomatically transformed rock is produced at the contact by hot solutions and vapours emanating from the intrusion. The rocks so formed are rich in distinctive lime-silicates such as idocrase, and the lime-rich garnets and pyroxenes (No. 279).

Mylonites are fine-grained, flinty, banded, or streaked out rocks, which have been produced as a result of the grinding and milling of rock masses along a thrust or fault zone. They are the end product of cataclastic deformation. They usually show 'eyes' of unaltered rock material, streaked out along the direction of movement. (Compare No. 246, deformed conglomerate, No. 243, augen-gneiss with No. 244, mylonite).

METEORITES

These are masses of matter from outer space, which have penetrated through the earth's atmosphere and reached the earth's surface. Falls of meteorites are a comparatively rare occurrence, less than 10 falls being recorded from the whole world in an average year, though a number of falls may have escaped observation. When matter, moving at very high velocity, reaches the earth's atmosphere, it becomes incandescent, to form a ball of fire. The meteorite soon loses its planetary velocity and explodes with a loud noise. Any fragments, which reach the earth's surface, are falling at only a few hundred feet a second. The stones from one fall at Hessle in Sweden in 1869 were seen to rebound from a few inches of ice. In a few localities, notably in Arizona and Siberia, large craters have been found, which are believed to be due to really large meteorites which have buried themselves deeply.

Meteorities are of two kinds. The *Stony Meteorites* (No. 267) have an oxidized outer crust and consist largely of pyroxene and olivine. They approach basaltic rocks in mineral composition, or more precisely as they are deficient in feldspar, the ultrabasic peridotites and pyroxenites. But these meteorites contain some native iron, which is of the rarest occurrence in terrestial rocks, and also other minerals which could not exist for long in the presence of oxygen and moisture and therefore are unkown in the rocks of the earth's crust. Stony meteorites are usually of quite small size. *Iron Meteorites* (Nos. 265-6) are made up of two different alloys of nickel and iron. On a polished surface the two alloys appear as light and dark bands lying parallel to the faces of an octahedron. Some specimens of iron meteorites are of great size. The explorer R. E. Peary found one mass 36½ tons in weight in Greenland and transported it back to the American Museum of Natural History.

FOR FURTHER READING

Hallam, A. *A Revolution in the Earth Sciences* 1973. Oxford U.P. 127 pp.

Harker, A. *Petrology for Students*, 8th Edn. 1954. Cambridge U.P. 283 pp.

Hatch, F. H. and Rastall, R. H. T*he Petrology of the Sedimentary Rocks* 5th Edn. revised by J. T. Greensmith 1971. Allen and Unwin, 502 pp.

Hatch, F. H., Wells, A. K. and Wells, M. K. *Petrology of the Igneous Rocks* 13th Edn. 1973. Allen and Unwin, 551 pp.

Heinrich, E. W. *Mineralogy and Geology of Radioactive Raw Materials*. 1958. McGraw Hill, 654 pp.

Jones W. R. and Williams, D. *Minerals and Mineral Deposits* 1954. Oxford U.P. 248 pp.

Kirkaldy, J.F. *General Principles of Geology* 3rd Edn. 1962. Hutchinson.

Read, H. G. *Rutley's Elements of Mineralogy* 25th Edn. 1963. Allen and Unwin, 525 pp.

Rittman, A. *Volcanoes and Their Activity* 1962. Interscience, 320 pp.

Smith, H. G. *Minerals and the Microscope* 4th Edn. revised by M. K. Wells 1956. Allen and Unwin, 148 pp.

Smith G. F. H. *Gemstones* 13th Edn. revised by F. C. Phillips 1958. Methuen, 560 pp.

GLOSSARY

Accessory Minerals Minerals present in small quantities in igneous and metamorphic rocks. Their presence or absence does not affect the name given to the rock.

Acid Rock Igneous rocks and gneisses containing a high proportion (approximately 65 % or above) of silica, shown by the presence of appreciable amounts of free quartz. These rocks are typically light in colour and relatively light in weight.

Alluvial Deposits The deposits, usually of sand or clay, laid down by rivers or in semi-arid regions by sheet floods.

Amorphous Complete absence of any crystalline structure and hence of any regular form.

Amygdales Almond-shaped cavities in lava flows formed by the escape of gases. The amygdales may remain empty or may later be lined with crystals or be completely infilled with zeolites or other secondary minerals.

Anticline An uparched structure of rocks, with the beds dipping away on either side from the axis.

Arenaceous Medium-grained sedimentary rocks which are dominantly made up of grains of sand grade, that is 0.1 to 2 mm. in diameter.

Argillaceous Fine-grained sedimentary rocks which are dominantly made up of particles of the clay grade, that is less than 0.1 mm. in diameter.

Assimilation The digestion of country rock by magma. The first stage is the fracturing of country rock by the intrusion of magma. Then the broken-off fragments (xenoliths) begin to lose their clear cut edges, until finally they are completely digested and disappear.

Atom The smallest part of a chemical element which can exist either alone or in combination with atoms of the same or of different elements.

Basic Rocks Igneous rocks and gneisses with no free quartz and composed mainly of dark-coloured minerals. The amount of silica present in such rocks is usually less than 55 %.

Batholith A large mass of plutonic igneous rock usually with a dome-shaped roof and steep walls. Batholiths were generally intruded during orogenic movements and tend to be elongated along the fold axes of the surrounding country rocks.

Bedding The layering visible in sedimentary and certain igneous rocks. The layers differ from one another in colour, coarseness of constituent material, etc. The layers represent successive positions of the upper surface of a growing pile of material.

Boss A relatively small and usually circular outcrop of plutonic igneous rock. Bosses are usually the highest (first to be exposed) parts of batholiths.

Calcareous Sedimentary rocks that are dominantly made up of calcite (calcium carbonate) and/or dolomite (calcium-magnesium carbonate).

Carbonacesous Sedimentary rocks that are dominantly made up of carbon or of carbon compounds.

Cement The material which binds together the particles of sedimentary rocks.

Clastic Sedimentary rocks which are dominantly made up of fragments of pre-existing rocks.

Cleavage Planes of easy splitting in minerals and pelitic metamorphic rocks. In minerals the cleavage planes are parallel to particular crystal faces, in the pelitic metamorphic rocks they are determined by the orientation of the mineral particles.

Contact Metamorphism The alteration of country rocks by intrusive igneous rocks. In the zone of altered rocks, the metamorphic aureole, new minerals are developed. These minerals, such as andalusite, cordierite, grossular and manganese garnet, are stable at high temperatures, but may be unstable at great pressures.

Crypto-crystalline Possessing only a trace of crystalline structure. This may only be visible under high magnification.

Crystalline A substance having an ordered atomic structure and a definite chemical composition.

Crystallized Minerals that occur as well developed crystals.

Cubic System The Crystal System possessing the highest degrees of symmetry. The three axes are all of equal length and are at right angles to one another. The forms of the Cubic System are all closed forms, e.g. cube, octahedron, etc. (see Fig. 11, I-VI.)

Diagenesis The physical and chemical changes which may take place in sediments during and after deposition, but before their consolidation.

Differentiation The processes by which originally homogeneous magma may split into fractions of different composition. One process (gravitational sorting) is the sinking of the heavier first formed crystals. These are of the more basic minerals, and therefore the remaining liquid becomes progressively more acidic.

Dissemination Minerals scattered irregularly throughout the host-rock.

Dyke A minor intrusion of igneous rock cutting the country rock at a high angle to its bedding. Most dykes are more or less vertical. They have a sheet-like form, being extremely thin as compared to their length.

Dynamic Metamorphism The alteration of pre-existing rocks by earth movements with the production of minerals, such as sillimanite and kyanite, which are stable at high pressures.

Eluvial Deposits Concentrations of minerals of high specific gravity as weathered material creeps downhill under gravity.

Era A major division, of the order of a hundred million years or more, of geological time.

Essential Mineral The minerals forming the major constituents of igneous and metamorphic rocks. The name given to such rocks is determined by the essential minerals present.

Evaporite Sedimentary rocks that are dominantly made up of the products of the evaporation of bodies of water.

Extrusive Rocks Igneous rocks which have consolidated from magma flowing either over the surface of the ground or into water. Owing to their rapid cooling, they are fine-grained or even glassy in texture.

Fault A fracture-plane in rocks, along which the rock-mass on the one side has been moved relative to the rock-mass on the other side.

Ferriferous Sedimentary rocks that are dominantly made up of compounds of iron. Such rocks are commonly worked as iron ores.

Fossil The remains, or traces, of organisms which inhabited the earth's surface during the geological past.

Gangue The non-metalliferous portion of the mineral infilling of a lode.

Geode A stone with a hollow interior into which project well formed crystals.

Geosyncline An elongated belt of the earth's crust in which a great thickness of sediments accumulated during long continental subsidence. The infilling of the geosyncline is later compressed and uplifted to produce a folded mountain chain.

Granular Composed of grains. This term may be applied to the form in which a mineral occurs or to the texture shown by metamorphic and some sedimentary and igneous rocks. Such rocks are made up of tightly interlocking, approximately equal-sized grains of one or more minerals.

Heavy Minerals Minerals normally present in very small quantities (less than one per cent) in sedimentary, particularly arenaceous, rocks. The specific gravity of heavy minerals is greater than that of quartz, so they can be concentrated by the use of heavy liquids, such as bromoform, on which the quartz will float. The study of the minerals which sink through the bromoform enable deductions to be made as to the nature of the rocks from which the sediment was derived.

Hexagonal System The Crystal System with a vertical and three equal horizontal axes set at 60° to each other. Hexagonal crystals are six- of twelve-sided in plan (see Fig. 12, I-V).

Hydrothermal Hot aqueous liquids at temperatures of about 500°C. or below, which are formed at a late stage in the consolidation of magma. These liquids may react with the country rock to produce hydrothermal alteration or they may be rich in metals and crystallize in fractures as mineral veins.

Hypabyssal Medium-textured, sometimes porphyritic, igneous rocks that crystallized fairly quickly at no great depth. They occur as minor intrusions, such as dykes and sills.

Igneous Rocks Rocks which have been formed from the consolidation of magma.

Intermediate Rocks Igneous rocks whose silica percentage is normally between 65 and 55 %. Free quartz, if present, is very subordinate in amount.

Joint A fracture-plane in rocks, along which there has been little or no relative movement of the rock masses on either side.

Lava Molten rock material (magma) which has risen up either a vent or a fissure and then spread across the earth's surface as a lava flow.

Leucocratic Minerals The light-coloured minerals (quartz, feldspar, etc.) of igneous and metamorphic rocks. A leucocratic rock is light in colour.

Lode Fissures and fractures in rocks infilled by metalliferous and non-metalliferous minerals of hydrothermal or pneumatolytic origin.

Mafic Minerals The dark-coloured minerals (olivine, pyroxene, amphibole, and mica) of igneous and metamorphic rocks. Melanocratic is a term applied to dark-coloured rocks with a high proportion of mafic minerals.

Magma The fluid raw material from which igneous rocks have consolidated. Magma is made up of molten rock material (a mixture of complex silicates and oxides) together with water vapour and other volatiles.

Matrix The fine-grained groundmass which cements together the coarser constituents of rocks.

Metamorphism The alteration of igneous and/or sedimentary rocks by earth movements and/or hot magma. The original rocks have been partially or completely reconstituted owing to the development of new structures, such as cleavage, and the formation of metamorphic minerals that are stable under conditions of considerable pressure and/or temperature.

Metasomatism The replacement of one mineral by another mineral owing to the introduction of material by percolating water or by some other external source. In this way the nature of the rock containing the original mineral may be greatly altered.

Mineral A natural inorganic substance with a definite chemical composition, atomic structure and physical properties.

Monoclinic System The Crystal System with three axes of unequal length. The vertical axis and left-right axes are at right angles, but the angle between the vertical and the front-rear axis is greater than a right angle. Monoclinic crystals are therefore not symmetrical in side view (see Fig. 11 IX).

Monomineralic A rock composed almost completely of only one mineral.

Ore A rock or mineral deposit from which one or more metals may be profitably extracted.

Orogenesis Mountain building caused by the intense lateral compression of a belt of the earth's crust.

Orthorhombic System The Crystal System with three axes of unequal length at right angles to each other. Orthorhombic crystals are rectangular in plan (see Fig. 11, VIII).

Oversaturated Rocks Igneous rocks containing free (uncombined) silica in the form of quartz.

Pelites Metamorphic or sedimentary rocks dominantly composed of particles of the clay grade (less than 0.1 mm. in diameter).

Period A unit of geological time, spanning many millions of years. Eras are subdivided into periods.

Phenocryst Large crystals, often of perfect crystal shape, occurring in igneous rocks, and set in a finer-grained groundmass.

Phosphatic Deposit Sedimentary rocks which are dominantly made up of calcium phosphate.

Placer Concentrations, often in economically valuable amounts, of minerals of high specific gravity and durability in eluvial or alluvial deposits.

Plutonic Coarse-grained igneous rocks which have crystallized slowly in large masses at considerable depths. Plutonic rocks form major intrusions such as batholiths.

Pneumatolysis The changes in mineralogical composition produced by gaseous emanations from a consolidating magma. Tourmaline is a typical pneumatolytic mineral.

Porphyroblast Large and often well formed crystals of one or more minerals in metamorphic rocks, set in a finer-grained groundmass of the same or different minerals. Texturally equivalent to the phenocrysts of igneous rocks.

Porphyry Medium-grained igneous rocks showing a porphyritic texture, that is with large crystals set in a fine-grained, or even a glassy, groundmass.

Psammites Metamorphic or sedimentary rocks dominantly composed of particles of the sand grade (0.1-2 mm. in diameter).

Pseudomorph A mineral assuming a form which does not belong to it and which is often typical of another mineral. Pseudomorphs can be formed by the slow alteration of one mineral to another or by the gradual molecular replacement of one mineral by another or by the infiltration of a mineral to fill a cavity which had originally been occupied by a different mineral.

Radioactive Mineral Minerals containing certain unstable elements whose atoms undergo spontaneous disintegration to form atoms of other elements.

Regional Metamorphism The alteration, both structurally and mineralogically, of pre-existing rocks by a combination of heat and pressure during orogenic (mountain building) movements. Regionally metamorphosed rocks are developed over great areas in the root regions of the older folded mountain chains, especially in the Shield areas of Pre-Cambrian rocks.

Residual Deposits The insoluble material left behind as rock material is subjected to strong chemical weathering in hot humid or semi-arid regions.

Rudaceous Coarse-grained sedimentary rocks which are dominantly made up of fragments greater than 2 mm. in diameter.

Saussuritization The alteration of plagioclase feldspar to white mica, zeolites or other secondary minerals.

Schillerization The presence in crystals of plagioclase feldspar of minute inclusions in parallel alignment. Produces a striking play of colours in hand-specimen.

Secondary Enrichment The deposition of mineral matter, including the formation of new minerals, around the level of the water table, the upper part of a mineral vein is subjected to weathering.

Secondary Mineral Minerals produced by the alteration or reconstruction of primary minerals or rocks.

Sedimentary Rocks Formed by the deposition, in the depressions of the earth's crust, of material worn away from pre-existing rocks.

Segregation Either the concentration in layers of heavy crystals which have formed at an early stage of the consolidation of magma or the squeezing out of the residual liquid between the growing crystals of a consolidating magma with the formation of pockets of ore deposits.

Series The rocks that were deposited during a particular era of geological time.

Siliceous Sedimentary rocks that are dominantly made up of silica of organic and/or chemical, as distinct from clastic, origin.

Sill A sheet-like minor intrusion of igneous rock that is sensibly concordant (parallel) with the stratification of the country-rock.

Stratigraphical Table The order of formation of the major stratified rock-units of the earth's crust.

Stockwork An interlacing network of small ore-bearing veins.

Symmetry The regularity in the arrangement of like faces and edges of crystals.

Syncline A downfolded structure of rocks with the beds dipping downwards on each side towards the axis.

System The rocks which were deposited during a particular period of geological time.

Tetragonal System The Crystal System with three axes at right angles. The two horizontal axes are of equal length, so that Tetragonal crystals are square or eight-sided in plan (see Fig. 11 VII).

Texture A term applied particularly to igneous rocks. The texture of a rock is determined by the size and shape of its constituent minerals and their arrangement relative to each other e.g. porphyritic texture, spherulitic texture, etc.

Thermal Metamorphism See Contact Metamorphism.

Thrust A sub-horizontal fracture-plane in rocks, along which the overlying rock-mass has been displaced horizontally or semi-horizontally relative to the rock-mass beneath.

Triclinic System The Crystal System with the lowest degree of symmetry. The three axes are all of unequal length and *are not* at right angles to each other.

Trigonal System The Crystal System with a vertical and three horizontal axes of equal length set at 60° to each other. The symmetry is lower than in the Hexagonal system, so Trigonal crystals are triangular in plan (see Fig. 12, VII).

Undersaturated Rocks Igneous rocks deficient in silica, so that the normal place of the feldspars is taken, partly or completely by the feldspathoid minerals (leucite, nepheline, etc.)

Ultrabasic Rocks Dark and heavy igneous rocks whose silica content is very low, often less than 50%. Such rocks are composed very largely of the dark-coloured silicate minerals (olivine, pyroxene or amphibole).

Vein See Lode

Vug A cavity in a lode lined by well formed crystals.

INDEX

Reference numbers to coloured plates are in bold type; figure nos. of line illustrations are in italics; page nos. of text in Roman.

ACTINOLITE **116**, 91, 101, 136-7

Agate **37**, 124

Agglomerate **164**, 87, 151

Alabaster 111, 133

Almandine 138, 139

Alnöite........ **179**, 154

Alum Shale **231**, 162

Amazonstone **86**, 112, 131

Amber **148-9**, 164

Amethyst.. **32**, 123, 126

Amphibole, **38**, 112-8, *23*, 114, 136-7

Amphibolite ... **240**, 173

Amygdaloidal habit . 101

Andalusite . **102**, 95, 100, 113, 140-1

Andesite **182-3**, 147, 152

Andradite 138, 139

Anhydrite . **112**, 128, 132

Anorthosite **220-1**,134,155

Anthophyllite **114**, 136, 137

Anticline *5*, 85, 92

Antimonite........ 123

Apatite **82-3**, *12*, 99, 102, 110, 130

Aplite **212**, 154

Apophyllite, **132**, 115, 142

Aquamarine ... **119**, 141

Aragonite . **73**, 110, 130

Arborescent habit *13*, 101

Arenaceous rocks 148, 157-9

Argeelité........... 117

Argillaceous rocks 148, 159-60

Arkose 157

Asbestos. **117-8**, 95, 137

Asphalt ... **158**, 148, 164

Asphaltite **158**, 145

Augen-gneiss... **243**, 169

Augite **109**, *23*, 100, 114, 135-6, 137, 147

Azurite.. **76**, *9*, 111, 131

BANDED ORES... **287**, 163

Banding **171**, **226**, **246**, **260-1**, 146

Barytes **77**, *11*, 91, 100, 101, 111, 132

Basalt . **184**, *27*, 147, 152

Bauxite........ 130, 164

Bedding, false.. *28*, 159

— graded...... *29*, 159

— plane *31*

— slumped *30*, 159

— true *28*, *30*, 159

Beef 130

Beryl **120-4**, 99, 102, 114, 126, 141

Biotite **138-9**, 81, 115, 138

Blende **21-2**, *7*, 96, 109, 122, 131

Bloodstone **40**, 124

Blue ground . **6**, 119, 156

Blue John ... **64**, 129

Bog iron ore **60-1**,127,162

Bone Bed...... **284**, 162

Bornite. **17**, *9*, 106, 121-2

Botryoidal 101

Boulder clay 157

Braunite.. **57**, 109, 127

Breccia.... **245**, 156, 165

Breccio-conglomerate 157

CALCAREOUS DEPOSITS 164-9

— rocks 148

Calcite **67-72**,*12*,91,99,100, 101, 110, 130, 165, 167

Carbonaceous deposits 148, 163-4

Carnelian **36**, 124

Carrara marble 173

Carstone 163

Cassiterite **46**, 95, 99, 108, 125

Cataclastic structure 169

Chalcedony .. **36-40**, 101, 107, 124

Chalcocite **18**,*9*,106, 121-2

Chalcopyrite **16**, *9*, 99, 106, 120, 121

Chalcotrichite 125

Chalk . **87**, 130, 148, 166

Chalybite . **74**, 111, 134

Chert 124, 148, 165

Chiastolite . *25*, 113, 140

Chinastone 129, 154, 167

Chlorite **142**,*41*,116,142-3

Chromite 91, 95

Chrysotile 137, 143

Cinnabar .. **271**, 106, 112

Citrin 123, 140

Clastic rocks **284**, 148, 156-62

Clay 148, 159, 160-1

Clay minerals........ 159

— rocks ... **230**, 159-60

Cleavage of minerals 102

—rocks **290**,*40*,89,169,170

Clinozoisite 141

Coal....... **238**, 148, 164

Cobaltite .. **26**, 107, 123

Coccoliths *36*, 116

Colour of minerals . 101

Concretions **11**, **12**, **237**, **290**, *39*, 159, 168

Condensed deposits 162

Conglomerate **224-5**, **246**, **283**, 118, 156, 165

Contact metasomatism *8*, 91, 101

Copper **4**,*9*,*13*,94,104,118

— pyrites **16**, 91, 106, 121

Coral 166

Cordierite **129-30**,89,115, 142

Corundum **47-9**, *19*, 102, 108, 126
Country rock *6, 7*, 88, 91, 122, 153, 154
Crocidolite. **38**, 124, 137
Cryolite ... **66**, 109, 129
Crystal axes ... 99-100
— form 96-102
— laltice *10*
— sorting 155
— systems 91-2
Crystallography.... 91
Crystals . *11, 12*, 96-102
Cube.............. *11*
Cubic system . *11*, 99-100
Cuprite...... *9*, 108, 125

DENDRITIC HABIT **274**, 101
Derbyshire spar..... 129
Diabase ... **187-90**, 155
Diamond **5, 6**, *11, 14*, 91, 99, 102, 104, 118-9
Diatomite 163
Diatoms *36*, 163
Diopside **108**, 135
Diorite...... 147, 154-5
Disthene....... 114, 141
Dodecahedron, pentogonal *11*
— rhombic *11*
Dolerite **187-90**, 147, 155
Dolomite **278**, *12*, 110, 131, 148, 167-8
Dolomitic Limestone *39* 167-8
Dome (oil bearing).. *33*

EARTH, CORE *1*, 82
— layered nature *1*, 81-3
— mantle..... *1*, 82, 91
Eclogite **263**, 156
Eleolite 113, 135
Emerald **120, 122**, 102, 126, 142
Emery 126
Enstatite... **107**, 114, ·135
Epidote **105-6**, 91, 114, 141
Euxenite... **154-5**, 114
Evaporite Deposits 88, 128, 132, 148, 162

FAULTING *4*, 85
Fault plane....... *4*, 92
— trap *33*
— zone 85
Feldspar, orthoclase **84-5**, 87, *11, 12*, 81, 99, 100, 102, 112, 133, 147
— plagioclase **88-9**, 99, 112, 134, 147
Feldspathoids 134-5, 147
Ferriferous rocks 148, 162-3
Filiform habit 101
Fissure eruptions ... 85
Flat 92
Flint **43-4**, 124, 6, 148, 163, 17
Flow structure . **171**, 146
Foliated habit 101
Footprints *31*, 151
Foraminifera *36*
Form in minerals ... 101, 104-16
Fossils......... 159, 162
Fuchsite **140-1**, 116, 138
Fullers' Earth . 161-2

GABBRO . **218-9**, 147, 155
Gadolinite ... **156**, 144-5
Galena **19-20**, *11*, 99, 106, 122, 131
Gangue minerals . 91, 92
Garnet **93-6**, *11, 24, 41*, 89, 91, 95, 99, 101, 113, 126, 138
Geological System .. 89
— Time Scale .. 89-91
Geosyncline. *1, 3*, 85, 89
Glauconite. **146**, 116, 143
Gneiss **255-61**, 89, 149, 172
Gold **1, 2**, 94-5, 104, 117
Gondwanaland *157*
Gossan...... *9*, 94, 127
Granite **172-6, 191-204**, 206, *3, 6*, 81, 85-6, 88, 147, ·53-4
— gneiss. **242**, 153, 172
Granular habit 101
Granulite .. **239**, 153, 170

Graphic granite. **213**, 154
Graphite **7**, 95, 104, 119
Greenstone......... 137
Greisening 154
Greywacke.. *29*, 158, 159
Grossularite 138
Gypsum **78-80**, *22*, 100, 101, 111, 119, 132-3, 148

HABITS OF MINERALS 101
Halite **63**, *10, 11, 20*, 109, 128-9
Hardness of minerals 101-2
Hauyne 135
Heavy minerals . 141, 157
Heavy Spar 111, 132
Hematite **50-1**, **273**, *18*, 108, 125
Hemidome......... *11*
Hertfordshire Pudding-stone **283**, 156
Heulandite **134**, 142
Hexagonal System *12*, 99
Hornblende **112-3**, *23*, 100, 136, 137, 147
Hornfels... **252**, 89, 172
Hyacinth........... 140
Hydrothermal solutions 91-2

ICELAND SPAR .. **67**, 130
Idocrase **98**, *11*, 113, 139
Igneous rocks *8*, 86, 91, 146-7, 150-6
Ilmenite **58**, 95, 109, 127
Intrusive rocks 138
Iron 104, 118, 122, 162-3
Iron pyrites **8-12**, 88, 91
Ironstone.......... 163

JADEITE 163
Jargoon 140
Jasper......... **39**, 124

KAOLINIZATION 154
Kidney iron ore 101, 108, 125
Kyanite **104**, *41*, 89, 95, 100, 101, 114, 141